A Legacy to Nuclear Science and
Engineering in Canada

Whiteshell Laboratories

Chris Saunders
Whiteshell History Committee

2017

First Printing: 2017

ISBN 978-0-9950984-1-1

Pinawa Foundation
Post Office Box 100
Pinawa, Manitoba R0E 1L0

Illustrations and photographs are courtesy of Atomic Energy of Canada Limited or from public sources unless otherwise indicated.

Contents

Foreword

This book, courtesy of the Whiteshell History Committee, weaves the story of Whiteshell Laboratories. Whiteshell research has focused on a wide range of subjects, including the building of the WR-1 test reactor, nuclear safety, environmental science, radiation chemistry, analytical science and nuclear waste management. Research contributions have also been made on topics such as small reactors, life sciences, commercializing science, environmental protection, and the decommissioning of nuclear facilities. Throughout the book we have included short biographies and pictures of some of the people that played important roles in establishing and operating the site.

Whiteshell's contribution to nuclear science and engineering in Canada is impressive. Although this book does not include all of the contributions that have been made over the past 50 years, it does illustrate the depth and breadth of the accomplishments that have made Whiteshell Laboratories an important part of the Canada's nuclear family.

The Whiteshell History Committee would like to thank the many people who contributed information, articles and photographs. Your contributions have helped make this book possible. We would also like to thank AECL for access to their archive of photographs and reports.

This book is dedicated to the thousands of men and women that have been part of the Whiteshell Story.

Abbreviations

ACR:	Advanced Candu Reactor
AEC:	Atomic Energy Commission (US)
AECB:	Atomic Energy Control Board; Now Called CNSC
AECL:	Atomic Energy of Canada Limited
ALWTC:	Active Liquid Waste Treatment Centre
ASB:	Analytic Science Branch
BCLT:	Buffer Coupon Long Term
BDA:	Blast Damage Assessment
CAD:	Computer Aided Drafting
CANDU:	Canada Deuterium Uranium
CATHENA:	Canadian Algorithm for Thermal Hydraulic Network Analysis
CCSF:	Concrete Canister Storage Facility
CCWESTT:	Canadian Coalition of Women in Engineering, Science, Trades and Technology
CNEA:	Canadian Nuclear Energy Alliance
CGE:	Canadian General Electric
CNL:	Canadian Nuclear Laboratories
CNSC:	Canadian Nuclear Safety Commission
COG:	CANDU Owners Group
CRL:	Chalk River Laboratories
CSE:	Composite Seal Experiment
CSSP:	Canadian Safeguards Support Program
CTB:	Chemical Technology Branch
CTF:	Containment Test Facility
CUPE:	Canadian Union of Public Employees
CVD:	Cerenkov Viewing Device
DCVD:	Digital Cerenkov Viewing Device
DHC:	Delayed Hydride Cracking
DIDO:	Research Reactor at Harwell, UK.
DNA:	Deoxyribonucleic Acid
EA:	Environmental Assessment
EDTA:	Ethylene Diamine Tetraacetic Acid
EIS:	Environmental Impact Statement
ENDRES:	Engineering Design of Repository Sealing Systems
ESR:	Electron Spin Resonance
FFS:	Fitness-For-Service
FIG:	Field Irradiation Gamma
HCF:	Hot Cell Facility
HEU:	Highly Enriched Uranium
IAEA:	International Atomic Energy Agency
IAFF:	International Association of Fire Fighters
IAM&AW:	International Association of Machinists and Aerospace Workers
IFTF:	Immobilized Fuel Test Facility
IRF:	Irradiation Research Facility
ITAP:	Isotope Technology Acceleration Program
LANL:	Los Alamos National Laboratory
LEU:	Low Enriched Fuels

LLRW:	Low-Level Radioactive Waste
LOCA:	Loss of Coolant Accident
LOEC	Loss of Emergency Coolant
LSVCTF:	Large Scale Vented Combustion Test Facility
MAPLE:	Multipurpose Applied Physics Lattice Experimental
MFR:	Moderately Fractured Rock
MLW:	Medium-Level Waste
NAA:	Neutron Activation Analysis
NDT:	Non-destructive Testing
NFWMA:	Nuclear Fuel Waste Management Agency
NFWMP:	Nuclear Fuel Waste Management Program
NISP:	Non-Reactor Isotope Supply Program
NLLP:	Nuclear Legacy Liabilities Program
NRC:	National Research Council
NRCan:	Natural Resources Canada
NPD:	Nuclear Power Demonstrator
NRU:	National Research Universal
NRX:	National Research Experiment
NSERC:	Natural Sciences and Engineering Research Council of Canada
NWMO:	Nuclear Waste Management Organization
NWS:	North Warning System
OCDRE:	Organic Cooled Deuterium Reactor Experiment
OCR:	Organic Cooled Reactor
OMRE:	Organic Moderated Research Experiment (Located in the US)
OTR:	Organic Test Reactor
PAR:	Passive Autocatalytic Recombiner
PCR:	Power Coefficient of Reactivity
PHT:	Primary Heat Transport
PHWR:	Pressurized Heavy Water Reactor
PIPE:	Prairies Isotope Production Enterprise
PIPSC:	Professional Institute of the Public Service of Canada
R&D:	Research and Development
RARB:	Radiation Application Research Branch
RERTR:	Reduced Enrichment for Research and Test Reactors
RIEX:	Recharge Infiltration Experiment
RMC:	Royal Military College of Canada
RTF:	Radioiodine Test Facility
SAP:	Sintered Aluminum Product
SCCG:	Sub-critical crack growth
SCWIST:	Society for Canadian Women In Science and Technology
SDR:	SLOWPOKE Demonstrator Reactor
SEAS:	Seal Evaluation and Assessment Studies
SEM:	Scanning Electron Microscope
SEU:	Slightly Enriched Uranium
SF:	Shielded Facility
SLOWPOKE:	Safe Low Power Critical Experiment
SMAGS:	Shielded Modular Above Ground Storage
SSC:	Soils Storage Compound
ST&I:	Science Technology and Innovation

SUGAR:	SLOWPOKE Uprated for General Applied Research
TIG:	Tungsten Inert Gas
TSX:	Tunnel Sealing Experiment
UA:	United Association of Plumbers, Pipe Fitters and Welders
UC:	Uranium Carbide
UKAEA:	United Kingdom Atomic Energy Authority
URL:	Underground Research Laboratory
US:	United States
USWA:	United Steel Workers of America
WIN:	Women in Nuclear
WinSETT:	Canadian Centre for Women in Science, Engineering, Trades and Technology
WL:	Whiteshell Laboratories
WLDP:	Whiteshell Laboratories Decommissioning Program
WMA:	Waste Management Area
WNRE:	Whiteshell Nuclear Research Establishment
WR-1:	Whiteshell Reactor - 1
XPS:	X-ray Photoelectron Spectroscopy
ZEEP:	Zero Energy Experimental Pile
ZEUS:	Zoological Environment Under Stress

Building Whiteshell Nuclear Research Establishment

Chapter 1
An Introduction

The Whiteshell Nuclear Research Establishment (WNRE) was a division of Atomic Energy of Canada Limited (AECL), and a sister laboratory to the Chalk River Laboratories (CRL), located at Chalk River, Ontario. First established in the early 1960s, WNRE was renamed Whiteshell Laboratories (WL) in the early 1980s. The management of both sites was transferred to Canadian Nuclear Laboratories (CNL) in 2014, with AECL remaining as the owner of the sites. Through its years of operation, Whiteshell Laboratories made significant contributions to the science and engineering knowledge of the day. Major development programs included the Whiteshell Reactor-1 (WR-1), the Nuclear Fuel Waste Management Program, the SLOWPOKE Demonstration Reactor, CANDU Reactor Safety research projects, and the Underground Research Laboratory (URL).

In the late 1990s, AECL began consolidating research and development activities at CRL and began preparations for decommissioning WL. Decommissioning included a staged shutdown of operations, planning documentation and licensing for decommissioning. As a prerequisite to AECL's application for a decommissioning licence, an environmental assessment (EA) was carried out according to Canadian environmental assessment legislation.

In 2002, the Canadian Nuclear Safety Commission (CNSC) issued the first decommissioning licence for WL. This six-year licence was the first decommissioning licence ever issued in Canada for a nuclear research establishment. Subsequent licence renewals have allowed the decommissioning activities to proceed continuously since 2002.

Although Whiteshell is being decommissioned, its technical achievements remain. This book describes the work of thousands of dedicated staff members who have contributed key technical achievements to the scientific and engineering legacy of Canada.

Whiteshell Nuclear Research Establishment

1

In the late 1950s scientists and engineers at CRL began examining alternate types of reactor heat transport systems, including heavy water and various organic compounds. Organic liquids could be used at higher temperatures than heavy water, and this would increase the power-generation efficiency. Other advantages included a lower price for the organic coolants and reduced corrosion in reactor circuits compared to aqueous coolants. An early pioneer, Ian MacKay, had considered organic heat transport systems when working at CRL in the early 1950s. The results of this work, along with the work of scientists from various other countries, were presented at the second United Nations Conference on the Peaceful Uses of Atomic Energy, held in September 1958. W.B. Lewis, AECL's President, was involved with the United Nations conferences and he proposed that CRL examine heat transport systems using organic coolants in greater depth.

At the inauguration of John Diefenbaker in 1958, the Prime Minister announced his 'Northern Vision,' a bold strategy to extend Canadian nationhood to the Arctic and develop its natural resources for the benefit of all Canadians. Dr. Lewis used Prime Minister Diefenbaker's "Northern Vision" strategy to promote AECL's reactor program using organic coolants.

By February 1959, AECL management had decided to recommend the organic-cooled experimental reactor to the board, with an eye to eventual use in the Canadian north. In March, the board approved a $500,000 study.

The idea of an organically cooled reactor (OCR) was not an entirely novel concept. There were two OCRs operating in the U.S. by the early 1960s. The Canadian OCR was intended to be a straightforward experimental project which, like its three predecessors, ZEEP, NRX, and NRU, would be located at CRL. An experimental loop was set up in NRU and NRX to test materials. Cabinet approved both the Organic-Cooled Deuterium-Reactor Experiment (OCDRE) and CANDU in mid-1959. By July, AECL suggested that, with a staff of 2,500 people, Chalk River was "near the saturation point for major facilities and we should be considering a new establishment."

A quick survey of federal research laboratories indicated that three provinces were sadly lacking: Newfoundland, Alberta and Manitoba. Newfoundland, it was felt, was not an option at the time. Alberta had no need of atomic energy, blessed as it was with abundant oil and gas. So it would be Manitoba.

AECL proposed that a new research laboratory be established in Manitoba to develop the OCDRE. A preliminary survey went forward under the supervision of Shawinigan Engineering. AECL president J.L.Gray journeyed to Manitoba to meet with premier Duff Roblin. On November 8, 1959, he reported progress to the board: a probable site near the Seven Sisters Falls on the Winnipeg River; and an opinion by the federal government's housing agency that a new town would be developed, using Winnipeg firms as required.

J.L. Gray

James Lorne Gray was born in Brandon, Manitoba in 1913. After public school in Winnipeg, he graduated with a Masters in Mechanical Engineering from the University of Saskatchewan in 1938. He joined the Royal Canadian Air Force in 1939.

Mr. Gray's scientific career began at the National Research Council in 1948. He was assigned to the "Chalk River" project. He advanced to become President of AECL in 1958. For the next 16 years he led the corporation through an impressive growth period that saw Canada become a leader in nuclear engineering and technologies.

Mr. Gray was appointed a Companion of the Order of Canada in 1969 and was awarded the Professional Engineers Gold Medal by the Association of Professional Engineers of Ontario in 1973.

The original estimates of OCDRE's cost hovered around $6.5 million. Site costs were minimal, since the reactor was to be in Chalk River. By September 1960, with moving costs included, it was $18 million. The story was a familiar one. The design was more complicated than originally allowed for, which upset the timing as well as the cost. There were difficulties in finding a proper organic coolant for the heat transport system. According to a later analysis; some areas of the design could only be guessed at, so that contingencies had to be increased to allow for the unknowns. Negotiations with Manitoba were also complicated. The new research centre would not be costless for the province. It would have to look after some of the infrastructure, such as roads and a bridge across the Winnipeg River, as well as housekeeping details. By the summer of 1960, agreement seemed far away. With help from the federal government, however, an agreement was approved by cabinet on July 21, 1960, and WNRE was born.

Mike Wright – Longest Serving
Whiteshell Site Head
1986 to 1998

The research centre and the town site were both to be on the edge of Whiteshell Provincial Park. A town site, later called Pinawa, was selected and designed by Central Mortgage and Housing. The first preference for a name was Whiteshell, but a postal address already existed elsewhere. The name was accepted after the board was told that Pinawa meant "slow, calm, gentle water". The initial target was 146 houses by 1963.

Final agreement was reached on joint facilities, between AECL and Manitoba, just as the company was reconsidering its commitment to OCDRE. Some members of the AECL board remained sceptical about organic reactors. And though Lewis remained interested in organics, he pointed out to the board that OCDRE was very much an unproven technology, while CANDU technology was far advanced.

The board awarded OCDRE a grudging go-ahead, provided it was kept in bounds and its costs under control. The board recommended to the cabinet that work proceed in stages, each new milestone depending on satisfactory completion of the preceding one. By the spring of 1961 the design still had several issues. Some AECL management raised the question of whether any alternative reactors should be considered for

Site Heads at Whiteshell Laboratories*

Ara Mooradian – 1965 to 1971
Archie Aikin – 1971 to 1973
Bob Hart – 1973 to 1978
Stan Hatcher – 1979 to 1982
Ralph Green – 1982 to 1986
Mike Wright – 1986 to 1998
Harry Johnson – 1990 to 1995**
Colin Allan – 1995 to 1998**
Cliff Zarecki – 1998 to 2004
Grant Koroll – 2005 to 2011
Randy Lovelace – 2011 to 2012
Russ Mellor – 2012 to 2014
Craig Michaluk – 2014 to 2015

* From 1961 to 1965, Fred Gilbert was responsible for the site construction and Bob Robertson lead the scientific team at WNRE.
** Harry Johnson was Mike Wright's Site Head designate at Whiteshell from 1990 to mid-1995; Colin Allan was the designate from mid-1995 to 1998.

construction at WNRE. Fred Gilbert, who had been appointed to manage Whiteshell, urged scrapping OCDRE and planning for an NRX-type reactor. An alternative plan was proposed. The idea kept the organic heat transport system but would resemble OCDRE, except a solid shield would be substituted with a pool tank. It was called simply OTR for Organic Test Reactor. A design would be ready for the start of the construction season in April 1962. It was this proposal the board accepted.

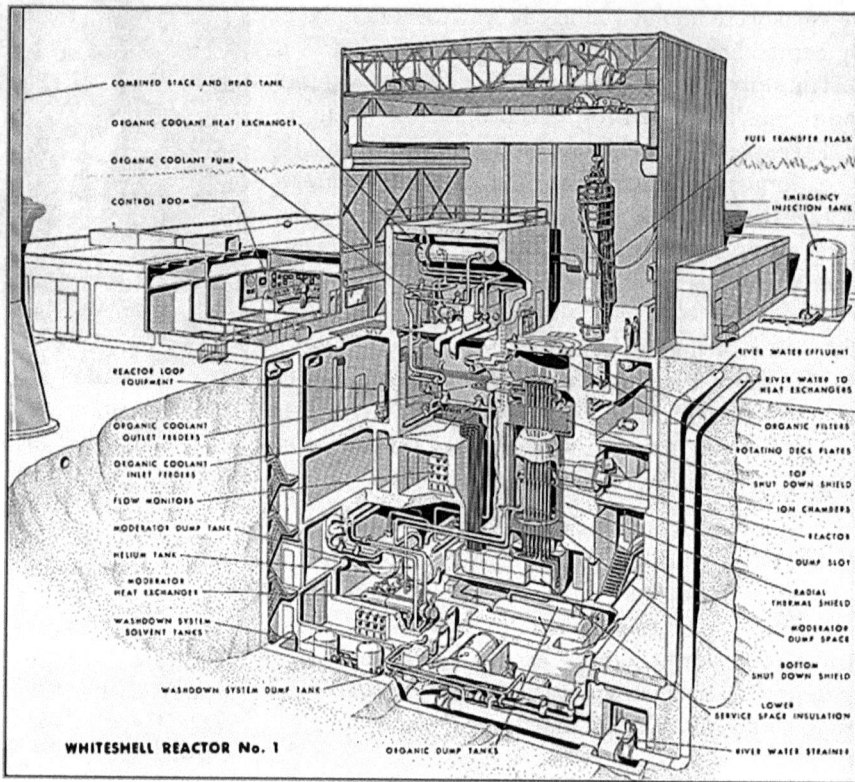

WHITESHELL REACTOR No. 1

Labels (clockwise): COMBINED STACK AND HEAD TANK, ORGANIC COOLANT HEAT EXCHANGER, ORGANIC COOLANT PUMP, CONTROL ROOM, FUEL TRANSFER FLASK, EMERGENCY INJECTION TANK, RIVER WATER EFFLUENT, RIVER WATER TO HEAT EXCHANGERS, ORGANIC FILTERS, ROTATING DECK PLATES, TOP SHUT DOWN SHIELD, ION CHAMBERS, REACTOR, DUMP SLOT, RADIAL THERMAL SHIELD, MODERATOR DUMP SPACE, BOTTOM SHUT DOWN SHIELD, LOWER SERVICE SPACE INSULATION, RIVER WATER STRAINER, ORGANIC DUMP TANKS, WASHDOWN SYSTEM DUMP TANK, WASHDOWN SYSTEM SOLVENT TANKS, MODERATOR HEAT EXCHANGER, HELIUM TANK, MODERATOR DUMP TANK, FLOW MONITORS, ORGANIC COOLANT INLET FEEDERS, ORGANIC COOLANT OUTLET FEEDERS, REACTOR LOOP EQUIPMENT

W.B. Lewis

Wilfrid Bennett Lewis, a physicist, dominated nuclear research and the development of nuclear power in Canada for nearly three decades. The development of the CANDU reactor was his most stunning achievement.

Born in England in 1908, Lewis earned a doctorate at Cavendish Laboratory in 1934 and continued his research on nuclear physics there until 1939. During the war he worked on the development of radar and in 1945 became superintendent of the Telecommunications Research Establishment at Malvern. A year later he agreed to head Canada's fledgling nuclear research facility at Chalk River, Ontario, where he made his professional home for the next twenty-seven years.

Lewis' drive, intelligence, and remarkable organizational skills placed him at the forefront of Canada's nuclear program, including establishing WNRE and Pinawa. Lewis fostered collaboration between AECL and Ontario Hydro that led to the development of the CANDU reactor. Lewis's influence on the development of science, technology, and industry in Canada and abroad was profound.

Do You Remember?

Mar 1940: George C. Laurence began experiments with a uranium-graphite sub-critical pile at the NRC in Ottawa.

Aug 17, 1942: C. D. Howe, Minister of Munitions and Supply, agreed to set up the Montréal Laboratory (forerunner of the CRL).

Jun 1943: Cominco produced Canada's first heavy water in Trail, B.C.

Jul 25, 1944: Sir John Cockcroft, CRL's first director, proposed the ZEEP project.

Aug 18, 1944: Chalk River site chosen.

Sep 5, 1945: ZEEP reactor attained first criticality.

Aug 31, 1946: Atomic Energy Control Act passed in Canada. This was the start of the Atomic Energy Control Board.

Sep 1946: W.B. Lewis became Director of the NRC's Atomic Energy Division, replacing Dr. Cockcroft.

Jul 22, 1947: The National Research Experimental reactor (NRX) at CRL attained first criticality.

Oct 31, 1947: The first isotope shipment was sent from NRX; Cerium-144 to the University of Saskatchewan.

Feb 14, 1952: AECL was incorporated as a federal crown corporation.

Apr 1, 1952: CRL was transferred from NRC to AECL.

Dec 12, 1952: An NRX accident leads to some fuel melting and fission product releases.

Feb 16, 1954: The second start-up of NRX was initiated following rebuilding.

Mar 27, 1957: AECL, Ontario Hydro and Canadian General Electric selected a horizontal pressure-tube design (CANDU).

Nov 3, 1957: The National Research Universal (NRU) at CRL attains first criticality.

May 7, 1958: J. L. Gray appointed President of AECL.

Oct 1, 1959: The decision was made to build the WNRE.

Oct 1962: A contract awarded to Canadian General Electric to build the WR-1 reactor.

Nov 1, 1965: WR-1 attained first criticality.

Dec 1965: Dr. Mooradian was appointed the first Managing Director of the WNRE.

Dec 16, 1965: WR-1 reached 100% power for the first time.

Feb 16, 1966: AECL's first labour strike occurred at WNRE involving Local 938 of the Canadian Union of Public Employees

Oct 1966: Two ponds were constructed near WL's WMA, to study the distribution of Cesium-137 received by plants and animals in and around the ponds.

*Jun 22, 1973:*W. B. Lewis retired from AECL.

Dec 1977: WR-1 was converted to uranium-carbide fuel.

Jan 24, 1978: KOSMOS-954 (USSR satellite with nuclear reactor) crashes in the North West Territories.

Feb. 5, 1978: The first shipment of radioactive space debris arrived at WL; *Operation Morning Light*.

1978: The Governments of Canada and Ontario established the Nuclear Fuel Waste Management Program (NFWMP).

Whiteshell Laboratories

1980: The Containment Test Facility (CTF) was built at WL to study hydrogen behaviour in containment.

Mar 1980: AECL received $40-million in funding to construct the URL.

May 26, 1984: Construction of the URL shaft began.

1985: URL opened.

1995: Phase 1 decommissioning of WR-1 and B100 was completed.

1997: AECL decided to discontinue most research programs and operations at WL.

1998: Both Acsion Industries and Ecomatters were formed to commercialize radiation-related science.

2001: A comprehensive environmental assessment to decommission the Whiteshell Laboratories was completed.

Apr 2. 2002: Federal Environment Minister David Anderson announced that the WL Decommissioning Project is not likely to cause significant adverse environmental effects, and that no further environmental assessment is warranted.

2002: The Federal Government concluded that deep geological disposal did not have the required level of acceptability to be adopted for managing nuclear fuel wastes.

Jun 13, 2002: Canadian Nuclear Fuel Waste Act was given royal assent.

Nov 15, 2002: The Nuclear Waste Management Organization (NWMO) is formed to secure public acceptance of selected nuclear waste disposal options.

Dec 19, 2002: CNSC announces 6-year decommissioning licence for WL. This is the first overall site decommissioning license ever issued in Canada.

Jun 2, 2006: NRCan announced $520 million funding over 5 years to begin the clean-up of the legacy waste at AECL's sites.

An early aerial photograph

Nov 17, 2010: URL was officially closed after 4 years of decommissioning.

2010: The Shielded Modular Above-Ground Storage Building and the Soil Storage Compound were constructed in the WMA at WL.

Sept 2015: AECL contracted Canadian National Energy Alliance to take over Canadian Nuclear Laboratories and complete the decommissioning of the WL.

Chapter 2
Early Years

The Whiteshell Reactor-1 (WR-1), WNRE's signature facility, was built starting in 1962. The 40 Megawatt reactor was designed and built by Canadian General Electric for $14.5 million. Shawinigan Engineering was responsible for constructing the building that housed the reactor. WR-1 was completed in June 1965.

WR-1 would test the concept of using an organic fluid to remove the heat generated by the reactor. Reactors using organic coolants can operate at higher temperatures and lower pressures than water-cooled reactors. Higher temperatures allow you to produce electricity with higher efficiencies. Lower pressures reduce maintenance costs and pressure vessel design requirements. WR-1 designers could use thinner-walled pressure tubes, giving the reactor a high neutron flux. WR-1 produced an average thermal neutron flux of $1.5 \times 10^{14}\,n/cm^2 s$.

WR-1 had vertical fuel channels. The neutrons were moderated by cool heavy water in a stainless steel calandria vessel (5 m high; 2.75 m diameter) surrounding the fuel channels. Fifty-four vertical aluminum tubes connected the ends of the calandria vessel. Pressure tubes, which contained the fuel and circulating organic coolant, were located inside the calandria tubes. The calandria tubes prevented the heavy water moderator from coming into contact with the pressure tubes. The fuel was compacted and sintered uranium dioxide, slightly enriched (2.4% U-235 in natural uranium), clad in a zirconium-2.5% niobium alloy.

The calandria vessel was divided into two sections. The upper section contained the fuel and the heavy water moderator. The lower section contained helium gas and collected the moderator spillage from the upper section. The reactor control system maintained the moderator level in the upper section by varying the helium pressure between the two calandria sections.

George C. Laurence

Born in Charlottetown, in 1905, Dr. Laurence's career spanned over 40 years. A student at the famous Cavendish Laboratory at Cambridge University from 1927 to 1930, Dr. Laurence studied alongside Ernest Rutherford, James Chadwick and John Cockcroft. He also worked at the National Research Council (NRC) on developing medical and industrial uses of radiology.

Dr. Laurence led the construction of a graphite-uranium nuclear assembly, becoming the first person in the world to induce fission by neutrons in an atomic pile. He directed groups developing instrumentation for ZEEP and the NRX and NRU reactors. He became Chairman of the Reactor Safety Advisory Committee in 1956 and President of the Atomic Energy Control Board from 1961 to 1970. Dr. Laurence spearheaded regulatory standards that are fundamental to Canada's nuclear industry.

Among his many honours, Dr. Laurence was appointed a Member of the Order of the British Empire and received the Medal for Achievement from the Canadian Association of Physicists, the Canadian Nuclear Society's W.B. Lewis Medal, and the American Nuclear Society's Certificate of Recognition.

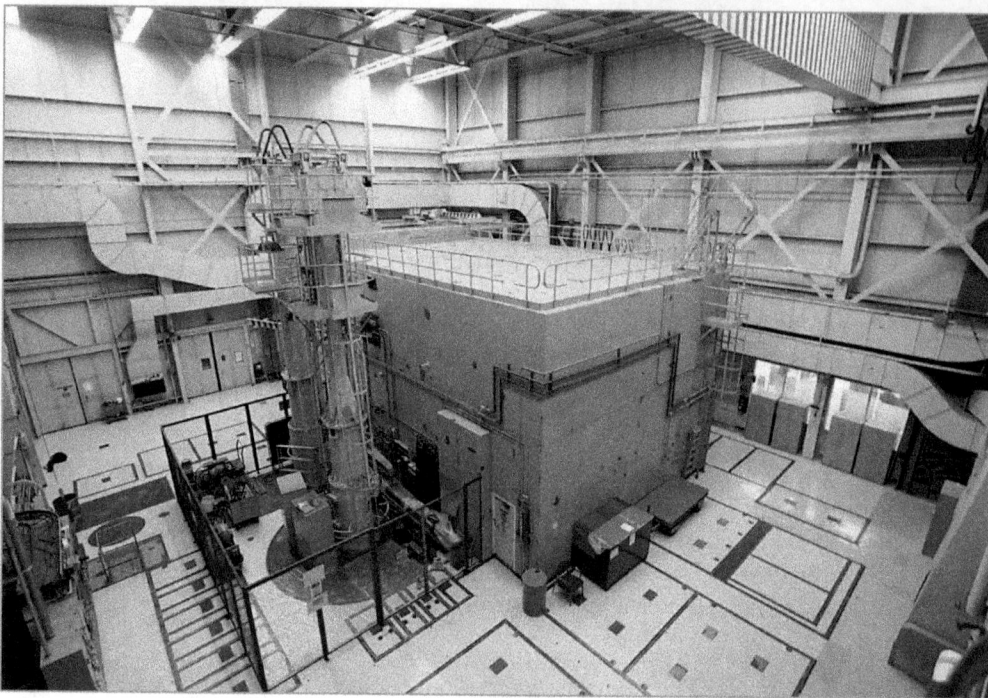

WR-1 Reactor

When the reactor was to be shut down, the helium gas pressure in the lower section was equalized with the upper section allowing the lower section to rapidly receive the moderator from the upper section. The moderator would then drain by gravity to a moderator storage tank.

The annuli between the fuel channels and the calandria tubes were purged with carbon dioxide gas to insulate the hot fuel channels from the moderator. Sampling of this gas provided a means of detecting moderator or organic coolant leaks between a fuel channel and a calandria tube.

The reactor core was surrounded by heavy concrete shielding (> 2 m thick). Stepped pipe chases through the concrete provided access for heavy water and helium lines and for the reactor vault exhaust duct. There were also three penetrations for the ion chambers. The inner surfaces of the concrete walls were cooled by embedded cooling coils.

The rotating deck plate provided an operational shield between the reactor and the reactor hall. The deck plate also supported the fuel transfer flask and provided the necessary radiation shielding during fuelling operations. The plates were comprised of cast steel (0.45 m thick) topped by wood fibre hardboard (Masonite; 9 cm thick) and a steel cover plate (0.5 cm thick). The deck plate had two holes for fuelling operations and periscope for viewing into the upper access space of the reactor.

The Primary Heat Transport (PHT) System removed the heat produced in the reactor core. The system was divided into three circuits. The removed heat was transferred to the Winnipeg River through three conventional tube-and-shell heat exchangers. River water was used as the secondary heat transfer coolant. The PHT system had three similar but independent circuits to achieve flexibility for experimental research.

To the outside world the most noticeable feature of WR-1 was the ventilation stack. The stack was known as the "stank" - a combination emergency coolant tank and ventilation stack.

During commissioning, some modifications were required in the pre-critical phase to make the various systems function as intended. For example, to prevent gas-locking of the moderator pumps during a moderator dump, it was necessary to extend the discharge pipe across the dump tank and install baffles to keep the entrained helium from the pump inlet. It was also necessary to re-route the ion-chamber cables to eliminate false signals from the crane control circuitry.

Building WR-1's Stank

WR-1 differed in several fundamental ways from NRX and NRU, besides having organic as the coolant. The reactor coolant could operate up to 425°C at the outlet, at the relatively low pressure of 2.15 MPa (g) at the inlet header. Another difference was that WR-1 had no control or shut-off rods. The moderator level and the addition of a soluble neutron absorber (boron) to the moderator controlled the power and shutdown functions.

Experimental Loops

A unique feature of WR-1 was its four in-reactor experimental loops and one out-of-reactor hydraulic test loop. Each in-reactor loop consisted of a fuelled test section and piping, equipment and instrumentation in an adjacent room to maintain the flow, pressure and temperature in the test section. A fuel position was converted to a loop by disconnecting the inlet and outlet feeders from the PHT and connecting the feeders to the loop inlet and outlet piping.

The out-of-reactor hydraulic test facility could handle full-sized fuel channels and fuel assemblies. The loop consisted of a circulation pump, a pressurizing pump, three test sections, three electric heaters, a make-up tank/degassifier, a condenser circuit, a purification circuit, a loop cooler, piping and instrumentation.

November 1, 1965 – Manitoba History

WR-1 went critical on November 1, 1965. The start-up was smooth and uneventful; the low-power commissioning continued throughout the month, with WR-1 operating almost continuously at 0.01% of full power. During the high-power commissioning of WR-1, there were difficulties with the automatic temperature control system, and the thermal power control system. Many of the components of this control circuits were replaced to ensure WR-1 operated properly.

WR-1 operators concluded that the commissioning of an organically cooled reactor was more straightforward than that of a pressurized-water-cooled system. They attributed this to two characteristics of the organic system: the high coolant-outlet temperature could be reached at a   lower operating pressure, and the radiation fields near the primary piping and headers were very low, allowing access to these areas during normal operation.

AECL's WR-1 Commissioning and Operations Team - Frank Oravec, Del Tegart, Jim Biggs, Bernie Gordon, Bernie Pannell, Vinny McCarthy, Grant Unsworth, Mickey Donnelly, Art Summach, Dick Meeker, Roy Barnsdale, Mike Berry, Al Nelson, Wilf Campbell, and Larry Gauthier

WR-1 Operator Bernie Pannell at the Controls, with Jim Biggs and Warner Brown on November 1, 1965

Jim Biggs, Bernie Pannell, John Weeks, and Del Tegart on November 1, 1965

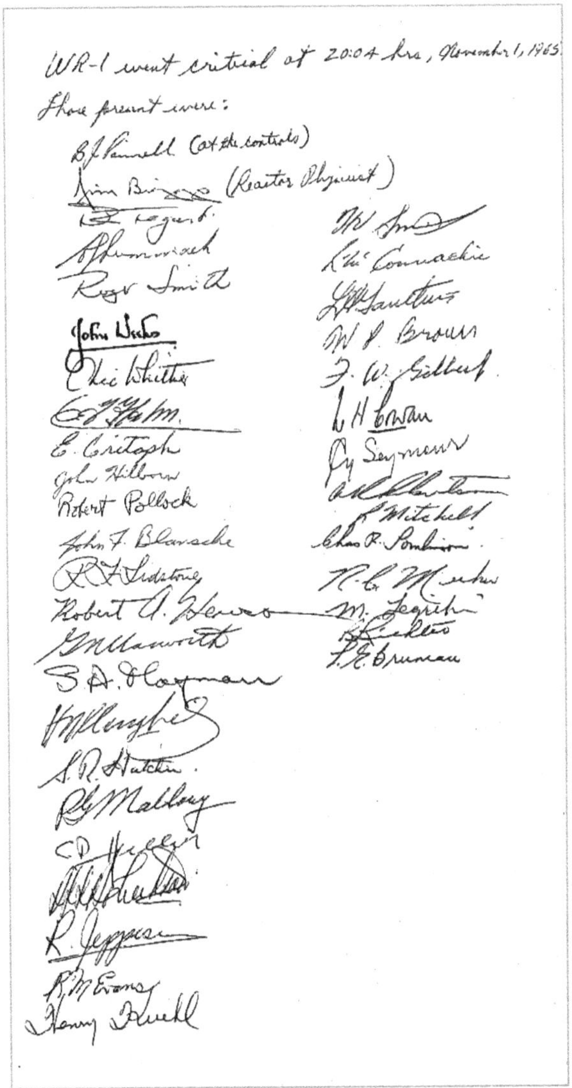

WR-1 went critical at 20:04 hrs, November 1, 1965.
Those present were:
B.J. Pannell (at the controls)
Jim Biggs (Reactor Physicist)

WR-1 went critical at 20:04 hrs November 1, 1965
Those present were:

B.J. (Bernie) Pannell
Jim G. Biggs
D.R. (Del) Tegart
A.J. (Art) Summach
Roger Smith
John Weeks

A.C. (Chic) Whittier, CGE
O.J. Holm, CGE
E. (Gene) Critoph
John Hilborn
Robert Pollock
John F. Blansche
Robert F. Lidstone
Robert A. (illegible)
G.N. (Grant) Unsworth
S.A. (Alex) Mayman
Norm Klingbeil, CGE
S.R. (Stan) Hatcher
P.G. Mallory, CGE
C.D. (Chet) Hillier

R.F.S. (Bob) Robertson
R. (Bob) Jeppeson

R.M. Evans
Henry Kuehl

Harry Smith

R. (Ross) McConnachie
L.H.W. (Larry) Gauthier
W.P. (Warner) Brown
F.W. (Fred) Gilbert
L.H. Cowan

Cy Seymour
A.R. (Alex) Robertson
R. (Ross) Mitchell
Chas. R. Tomlinson
R.C. (Dick) Meeker
M. (Manfred) Legiehn
B. (Ben) Richter
L.G. (Louis) Bruneau

Recording History – Signatures of witnesses to WR-1's first run-up to criticality on November 1, 1965.

WR-1 Research

Over the next four years WR-1 operated well, with an average availability factor of 85 percent. As with their commissioning experience, the operators concluded that the WR-1 operation was more trouble-free than that of a pressurized-water-cooled system. One issue with the organic coolant was its flammability. Fires were prevented by explosion proof wiring, good housekeeping, eliminating ignition sources by using improved pipe fittings, grounding all piping, and protection of areas containing organic circuits with fog and water sprays.

From 1966 to 1969, the operation of WR-1 was studied extensively. Staff demonstrated that fouling of heat transfer surfaces and the hydrogen uptake by zirconium alloys in the PHT system could be controlled by adjusting the coolant chemistry. The coolant had to be free of chlorine and low in metal oxides. Chlorine was removed by having a small amount of water in the coolant and by side-stream circulation through clay columns. This also produced an oxide film on the zirconium as a barrier to hydrogen migration and hydride formation.

WR-1 Fuel Transfer Flask

With WR-1 operating well, there was considerable incentive to upgrade the facility to expand the type of experiments that could be done and to lower the fueling costs. A number of changes were made in 1975. The stainless-steel pressure tubes were replaced by zirconium-alloy tubes to reduce neutron absorption. A third organic primary coolant circuit was added to service seventeen of the existing but as yet unused fuel sites. This third circuit increased the capacity of WR-1 to 60 MW.

Ara Mooradian

Born in 1922 in Hamilton, Ontario, Ara Mooradian gained his early training as an engineer and scientist at the University of Saskatchewan and the University of Missouri. His career began at the Consolidated Mining & Smelting Company before joining the staff at the Chalk River Nuclear Laboratories. At Chalk River, he was Head of the Development Engineering and Fuel Development Branches.

In 1965 Ara became the first Managing Director of the WNRE. He became Vice-President of the Chalk River Nuclear Laboratories in 1971 before taking up the position of Executive Vice-President for Research & Development at AECL (1977) and then Corporate Vice-President (1978). Dr. Mooradian was noted for his contributions to the development of low cost fuel for CANDU reactors. His honours have included the Canada Medal, the W.B. Lewis Award and Fellowships of the Royal Society of Canada and the Chemical Institute of Canada. He was also the first Mayor of the Town of Deep River.

WR-1 Fuel

Fuel development work for both WR-1 and CANDU-OCR concepts was carried out in the X-7 and U-3 loops at CRL. The initial driver fuel used in WR-1 was of an eighteen element uranium-dioxide design similar to the Douglas Point nineteen-element bundle, except that the centre element was omitted so that the bundle would fit over a central support shaft. The uranium was enriched in the range 1.2 to 2.4 % uranium-235.

Two fuel types were used:

- A Zr-2.5Nb-sheathed fuel element assembled into 49.5 cm long bundles, with each reactor fuel string having five bundles suspended on the central hanger tube.
- A sintered aluminum sheathed fuel element assembled into 81.3 cm long bundles, with each reactor-fuel string having three bundles suspended on the central hanger tube.

Dick Meeker and Phil Roy controlling the WR-1 Reactor

WR-1 operated with the uranium-dioxide driver fuel from start-up in 1965 until 1973. Experience with the uranium-dioxide fuel in WR-1 was not good compared to the performance of uranium dioxide in water-cooled reactors. The number of fuel strings removed due to failure or sticking exceeded those retired without incident. The overall average burn-up for all the uranium-dioxide fuel irradiated in WR-1 was 128 MWh/kgU, which satisfied the original target of 120 MWh/kgU.

Experimental irradiations of uranium-carbide fuel started in 1966 and in 1973 it was decided to use uranium-carbide as the reactor-driver fuel.

Conversion to uranium-carbide fuel was completed by the end of 1977. The performance of the uranium-carbide fuel was excellent. The average burn-up of the first 125 retired bundles was 253 MWh/ kgU (original target burn-up was 240 MWh/ kgU). The failure frequency of the fuel was also low and the consequences of failure were not serious. While activity releases were high enough to detect failure, they remained low enough to continue operation of the failed fuel until the next scheduled shutdown. One identified problem was hydrogen migration to the bundle end plates that caused

Frank Oravec and Alex Robertson in WR-1 Control Room

their embrittlement. Six bundles had to be retired due to end-plate breakage during shuffling operations. The problem was solved by using larger hydrogen-sink volumes in the end-plate region.

During the 20 years of WR-1 operation, there were four significant incidents:

- The failure of a large isolating valve on the cooling water line feeding the main Heat Exchanger for the "A" Circuit, resulting in the flooding of the lower level of the reactor building with river water.

- The failure of experimental fuel bundle 913 during reactor start-up that released sufficient gamma emitting fission products into the coolant system that the reactor was shut down for 6 weeks to remove the contaminants.

- The failure of a tube in one of the main heat exchangers resulted in the leakage of about 2,000 litres of organic coolant into the Winnipeg River. The organic coolant concentration and radiation levels in the river diminished to negligible levels in a short distance from the point of release.

- A mechanical seal failure of one of the main circulation pumps, increasing the risk of a loss of coolant.

WR-1: Unique Among Research Reactors

WR-1 was a useful research facility, testing experimental fuels, reactor materials, and other coolants for 20 years. It was the centre piece of a thriving research community - 567 employees by 1967. The reactor was a busy place, usually working around the clock. It had an availability of 85% over its lifetime, which was exceptionally high for a research reactor.

AECL designed a full-scale organic cooled prototype power reactor for evaluation by Ontario Hydro, based in part on the work at Whiteshell. The evaluation indicated that an OCR could be built for 10% less and with an annual operating cost of 10% less than the existing

Loading Reactor Fuel – Blake Cutting and Glen Snider

PHWR. OCRs were also shown to have low corrosion rates and reduced activity transport in the primary circuit. Ontario Hydro concluded that the logistics associated with training operators and developing the resources to supply equipment and construct an OCR would likely result in additional costs exceeding these expected capital and operating cost savings.

WR-1 was shut down for the last time on May 17, 1985, its place in history secured as the world's only operating heavy water-moderated reactor cooled by an organic fluid. The reactor has been defueled and largely disassembled. The reactor is now in an interim decommissioning stage, waiting final decommissioning.

The legacy of WR-1 is that organic coolants are still being considered for future reactor designs. Higher operating temperatures increase the thermal efficiency of organic power reactors. Lower pressures reduce maintenance costs and pressure vessel design requirements. Design features of

WR-1 Calandria

WR-1 have found their way into the nuclear reactors of today; instrument triplication, parameter duplication and frequent testing; transfer systems to move irradiated fuel, fuel channels and equipment safely from the reactor to water-filled storage facilities; and interim dry storage of fuel.

Archie Aikin

Dr. Archie Aikin was born in Saskatoon and attended schools in Winnipeg and Montreal. He obtained a B.Sc. (honours chemistry) from McGill University in 1941, served in the Canadian Army, then returned to McGill to gain his PhD. in chemistry.

Dr. Aikin joined the staff at Chalk River Nuclear Laboratories in 1949 and served there in a series of positions that included work in chemical engineering, nuclear fuels and economic evaluations of nuclear power systems. In 1968, he was appointed to Head Office, Ottawa, to set up the nuclear power marketing section as general manager. He was appointed vice-president, Whiteshell Nuclear Research Establishment, on January 1, 1971. He moved to be Vice-President, Commercial Products, for AECL in 1974.

M.D. Ward and R.E. Barnsdale adjusting guide ring on WR-1 fuel string – 1965

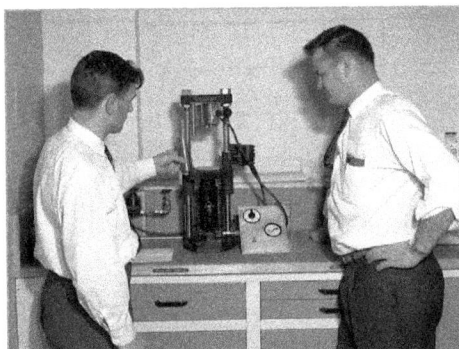

Doug Badger and Ray Sochaski operating automatic specimen mounting press – 1966

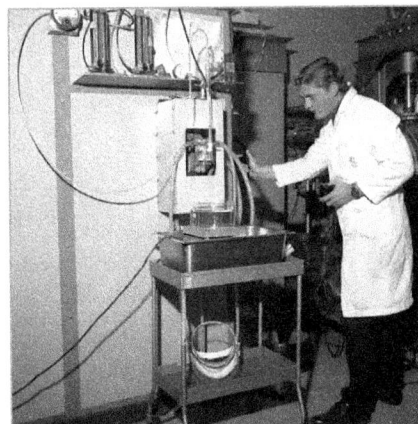

Lorne Swanson quenching molten uranium carbide in water - 1971

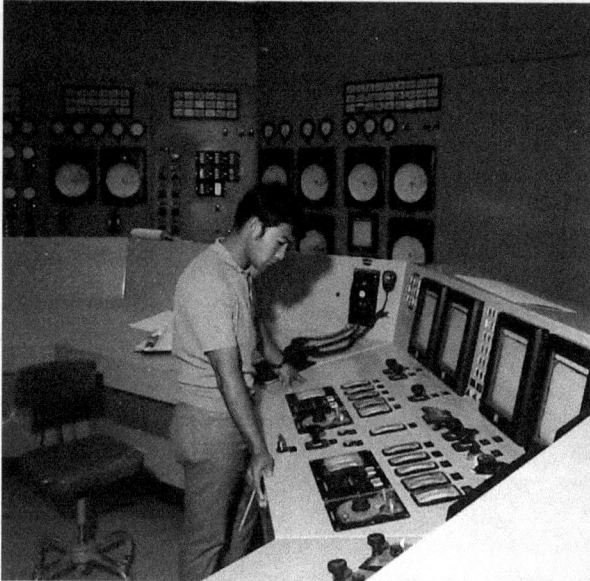

Mitch Ohta at WR-1 Operations

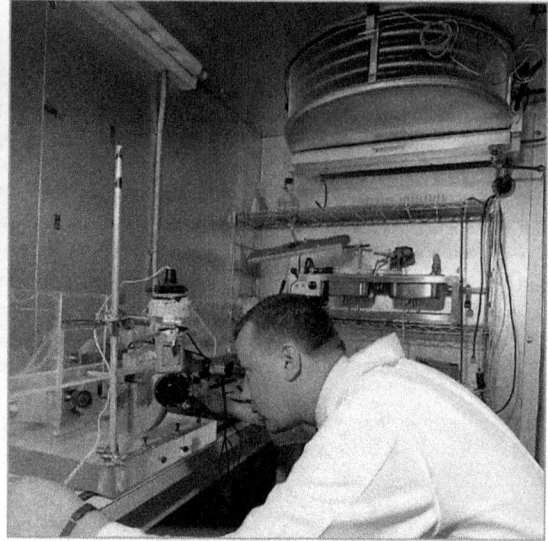

Bill Chelack observing membrane formation

L.R. Haywood presenting F.W. Gilbert award for 1 million hours accident free at
WNRE – 1965; with G. Nayler and J.L. Weeks

Life and Environmental Sciences

Whiteshell research in life sciences focused on the biological effects of radiation, the behaviour of radionuclides in the environment, the use of radionuclides for biological research, and developing instruments for radiation protection.

Many studies have been carried out in various countries, including Canada, on workers who were occupationally exposed to whole-body radiation, and on populations who live in areas where natural background radiation is higher than average. Those studies have been inconclusive; if there is any effect, it is too small to be detected against the normal high background of cancers attributable to other causes. Work at Whiteshell greatly improved our understanding of how radiation produces biological changes in living organisms and how best to protect nuclear workers and the general public.

Stan Hatcher, Roger Smith, Ara Mooradian, Bob Hart, Art Summach; Standing, Peter Dyne, John Weeks, Doug Molnar and Jim Putnam

Environmental Protection

From the earliest days of nuclear development there has been concern with the protection of workers and the general public, along with the movement of radioactive material through the environment, its uptake by plants and animals and the effects upon them. In the philosophy of radiation protection, it has long been believed that if people are protected, then the survival of other species in the environment will also be ensured. This topic remains a key component of nuclear environmental assessments worldwide, and early work from WL is often cited.

One of the first steps in commissioning the WNRE site was to undertake a preoperational survey in surrounding areas. The survey determined existing background radioactivity, the type and abundance of plant and animal life, and the dilution capacity of the Winnipeg River. AECL established a network of monitoring stations around the site and collected samples of fish and other animals to determine the content of radioactive materials.

It was important to obtain experimental data on the dilution and dispersion of radioactive material released into the atmosphere, soil and bodies of water. Since some of these

Peter Dyne, Bob Robertson, John Weeks and Abe Petkau

bodies of water were public areas, fluorescent dyes rather than radioactive materials were used as tracers. Staff released concentrated solutions of fluorescent dyes in the Winnipeg River at Whiteshell and, at different times after the release, measured the concentrations of dye downstream in using a fluorometer mounted in a boat. The data were used to estimate physical dilution of effluents from WNRE once it was operating.

Routine environmental surveillance at WL began in 1962. As part of the radiation safety program, radiation physics programs were developed, dosimetry services were offered, and industrial safety and radiation worker training courses were developed. A whole body counter was installed at Whiteshell and routine radiation surveying at specific facilities became mandatory.

Early environmental monitoring equipment

Early work determined that the general movement of groundwater in the Waste Management Area (WMA) was upward. Any releases to the groundwater would take hundreds to thousands of years to reach the Winnipeg River. Hydrogeologic studies were conducted to predict the subsurface behavior of radioactive contaminants (Cherry et al., Canadian Geotechnical Journal, 12 (1), 1973).

John Weeks –WL Pioneer in Environmental Monitoring

Orville Acres collecting air samples

Alex Henschell: A career trapper, hired by AECL to help acquire, prepare and establish radioactive characteristics of local animals, soils and plants. (Photograph courtesy of Ray Henshell)

Ab Reimer measuring atmospheric parameters

The WMA is composed of deposits of clay till above a sandy aquifer which overlies the bedrock. The bedrock is 15 m below ground surface. The hydrogeological investigations lasted 5 years and involved three main parameters:

1. expected residence times of radionuclides which may enter the groundwater flow system,
2. anticipated travel paths and discharge processes, and
3. suitability of the hydrogeological environment for physical manipulation to achieve greater containment capabilities.

The studies included test drilling, mapping of hydraulic head distributions using wells and piezometers, field permeability tests using single well response tests and long- and short-term pumping tests, mapping of natural hydrochemical patterns in the groundwater zone, tritium tracer experiments, groundwater age dating using ^{14}C and mathematical modeling using digital-simulation programs. Comparisons of the results indicated that there was a reasonable level of predictability of the hydrogeological environment in the area.

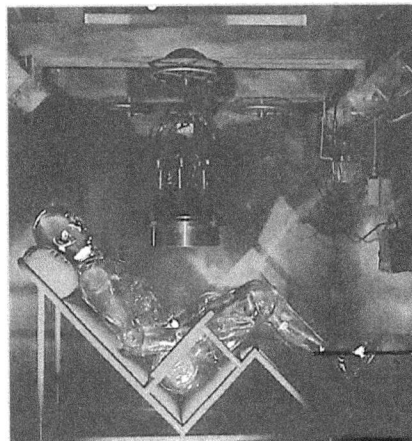

Whole Body Counter

The meteorology program focused on developing a model of atmospheric dispersion to be used in the case of an accidental airborne release.

Routine surveillance carried out on samples of water, air, fish, plants and soil ensured that releases of radioactive materials were below specified limits and acted as checks on the models that were being developed. Aspects of that work continue today as part of the decommissioning.

Whiteshell was a world leader in environmental protection, using the many unique facilities at the laboratory, such as the Cesium Pond, the Field-Irradiator Gamma (FIG) area, and the Zoological Environment Under Stress (ZEUS) area, to examine how radiation interacted with the environment.

Caroline Brady

Darlene Hood

Cesium Pond Experiments

In October 1966, two ponds were constructed northeast of the WMA, near the edge of a stand of aspen. The purpose of the ponds was to study the distribution of dose received by organisms living at the mud-water interface (References: AECL-3463 & Canadian Journal of Zoology, V.47, N.1, 1969, pp 17-21). The ponds were constructed by excavating 19-m diameter circular depressions. The depressions for the ponds were sloped to a maximum depth of 2 m at the centre and then lined with plastic sheeting. After the thaw in 1967, both ponds were in filled with melt water and spring rain. The pond areas were allowed to develop naturally. A total of 18.5 GBq (0.5 Ci) of $Cs^{137}Cl$ was injected into the east pond in 1967. The other pond was left in its natural state to serve as a control pond.

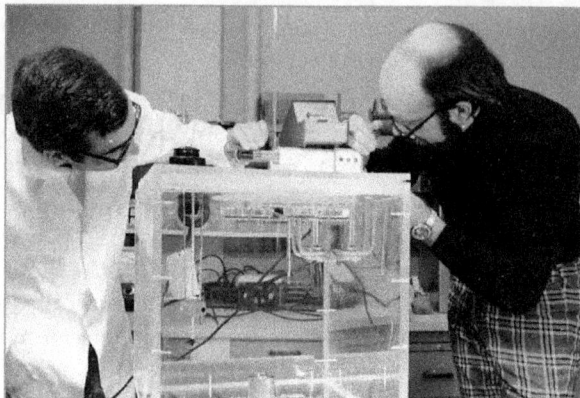
Environmental science at work

Research over the years focused of topics such as:

- How to use radionuclides as ecological tracers to study trophic-level relationships (J. E. Guthrie, et al., Canadian Entomologist, Vol. 101, Issue 08, August 1969, pp 856-861). To use cesium-137 to investigate the role of the giant water bug, *Lethocerus americanus* (Leidy), in the food-web of a small pond, it was necessary to measure the elimination rate of the nuclide.

- Effects of Cesium-137 on the pond ecosystems (Dugle et al., AECL-3463, 1970). The algal populations were examined as a function of Cs-137 exposure time.

- Radioecology, the study of the relationship of radioactive substances and ionizing radiation with the environment (The Manitoba Entomologist Vol. 4, 1970). One aspect of this study was the uptake and distribution of radionuclides such as Cs-137 by insects to study the significance of ionizing radiation as an ecological stress.

Keith Dalby, Jock Guthrie, and Murray Smith

Ethel Reich - Radioisotope Analysis

Over the years, the soil contamination extended approximately 10 m from the original pond boundary. The majority of the activity was located in the surface of the soil within 1 m of the ponds edge. All of the contaminated soil has now been collected and moved to the WMA for storage as part of site decommissioning.

Field Irradiator Gamma (FIG)

The Field-Irradiator Gamma (FIG) experiment irradiated a section of the Canadian boreal forest from 1973 to 1986. The radiation point source was 370 TBq of Cs-137 placed at a height of 20 m. The dose rates ranged from 0.12 to 1560 mGy/d. Effects of radiation were investigated for the tree canopy, natural growing shrubs, ground cover species, germination of seeds, morphological change and tree-ring growth.

Jan Dugle and a summer student collecting samples

Irradiation resulted in the establishment of four zones of vegetation: an herbaceous community (up to 65 mGy/h), a shrub community (0.1 and 1 mGy/h), a narrow zone of dying trees (>2 Gy/h), and a zone with no apparent impacts (<0.1 mGy/h). Concentrations of ^{14}C, ^{99}Tc, ^{129}I, ^{137}Cs and ^{226}Ra that could cause a dose rate of 0.1 mGy/h within vegetation were calculated. Chemically toxic effects of ^{99}Tc and ^{129}I would occur before radiological effects are predicted to occur. The calculated ^{226}Ra concentration was about a factor of 10 greater than that measured at some natural sites. Another experimental observation was that the seed germination of Jack Pine showed deleterious effects at 1.1 mGy/h. In contrast, the same work reported hormetic effects at dose rates of up to 0.6 mGy/h.

Experiments showed that tree swallows (*Iridoprocne bicolor*) and house wrens (*Troglodytes aedon*) avoided nesting in areas of high radiation (Zach and Mayoh, Ecology, Volume: 63:6, 1982). It appeared that the birds responded to radiation levels as low as 100 times background but it was not clear whether they detected radiation or simply responded to secondary clues.

The number of swallows and wrens fledged per box was unrelated to radiation exposure. The same was true for number of eggs, hatching success, fledging

Field Irradiator Gamma

success, incubation time, and nestling time. Breeding success was reduced because of infertile eggs, eggs with dead embryos, cracked eggs, predation, adverse weather, abandonment, and parasites. The logistic model was ideally suited for describing gains in mass in nestling swallows and wrens. The data showed that birds avoided adverse effects of radiation by judicious box selection. However, there were indications that at higher breeding densities birds may use high-exposure boxes, where breeding success or growth of nestlings may become reduced due to radiation.

Although physiological effects similar to hormesis were demonstrated at low levels of exposure in meadow voles (Mihok, Schwartz, & Iverson, Annales Zoologica Fennici, Volume 22, 1985), no effects on population numbers, survival, or reproduction were found up to an average lifetime dose of about 5.7 Gy, at a measured dose rate of 44 mGy/d. These results were similar to what was found during long-term monitoring of small mammal communities living in a gradient of gamma radiation over a large area in the FIG experiment.

Zoological Environment Under Stress (ZEUS)

The ZEUS project involved the chronic irradiation of meadow voles at nominal exposure rates of up to 40,000 times background. It remains one of the few well-controlled studies ever conducted over multiple generations of animals under natural conditions. The ZEUS experiments remain unique and are of current interest as a comparison to results of studies in the contaminated areas surrounding Chernobyl.

The facility was established in deciduous forest in 1974. It consisted of six 1-ha island grassland grids (100 m by 100 m) that were cleared of trees and seeded with grasses and clovers to provide habitat for voles, which rapidly colonized the area. The grids were set up in an array with each grid separated from the next by 100 m of forest. Grid 1 was the permanent control area. Wild meadow voles were exposed to gamma radiation from a Cs-137 field irradiator on a rotating basis.

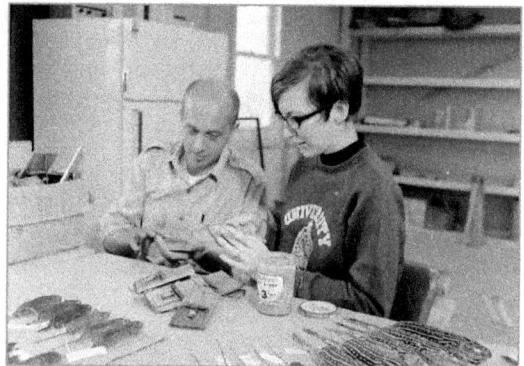

Stu Iverson and a summer student examining meadow voles

Irradiation was continuous during each experimental period, with interruptions required only to provide access to the grid. Vole demography and health were monitored by live-trapping throughout the year. Experiments were started in November, after the end of reproduction. This timing ensured that the starting generation consisted of a reasonably homogeneous group of late season young-of-the-year, with few early-season voles present, and likely no overwintered voles from the previous year still alive. This parental generation was then irradiated for about six months before breeding commenced the next spring. Samples of frozen blood plasma collected in association with routine diagnostic hematology were also available for glucocorticoid analysis and genetic studies.

No catastrophic effects were observed in ZEUS. The conclusions were that animals are stressed at very low doses and this can be measured. Short-lived prolific small mammals are nominally healthy when exposed to doses of 50 mGy/d of nearly continuous gamma exposure. Low, chronic doses of gamma radiation at 50 to 200 times background levels had beneficial effects on the stress axis and

ZEUS Experiment

the immune axis of natural populations of meadow voles (Boonstra et al., Environmental Toxicology and Chemistry, 2005). This work provided the first evidence of hormesis on mammals from the only large-scale, long-term experimental field test.

Chapter 3
Nuclear Mission

The safety of nuclear reactors has been a matter of major interest from the beginning of Canada's nuclear program. CANDU designers and researchers realized that information would be needed on how the various plant systems would behave, both at normal operating levels and during various postulated accidents. They also realized that the regulatory agency would require accurate information on the safety margins and the possible consequences of various accident scenarios.

In the early days of the nuclear power program (the late fifties), safety analysis dealt mainly with the consequences of the failure of individual critical process systems and/or pieces of equipment, as identified during the design process. However, as the program developed and the plants became larger, more attention was paid to dual failures, situations that could occur much less frequently but could have much greater consequences.

Robert Hart

Robert Hart arrived at CRL in 1948. After working on various projects including purification of heavy water, reprocessing of nuclear fuels, and studying the physical properties of these fuels, he moved to Whiteshell in 1965 as head of the Reactor Core Technology Branch.

He was appointed director of the Applied Science Division in 1969, managing director of the Whiteshell Site in 1973 and a vice president of AECL in 1974. In 1978, Bob became executive vice-president in charge of the AECL Research Company. Bob was awarded the W.B. Lewis medal by the Canadian Nuclear Association in 1981, with the following citation: "For giving the Whiteshell Nuclear Research Establishment world recognition in such fields as organic heat transport technology, thermalhydraulic technology for nuclear safety technology analysis, radioactive waste management".

Reactor safety programs all over the world expanded greatly in the 1970s. These programs concentrated more and more on investigating the consequences of low-frequency, worst-case accident scenarios. This thrust for more detailed safety analyses tended to be self-ratcheting; i.e., the more detailed the model, the more questions that were raised, requiring more research to answer them.

From the outset, the objectives of the CANDU reactor safety program were to develop a thorough understanding of the phenomena that might occur during reactor accidents, and to develop and verify the mathematical simulations used in the plant safety analyses.

The safety programs divided naturally into several areas:

- reactor physics, concerned with the reactivity effects associated with loss-of-coolant accidents (LOCA);

- thermal hydraulics, dealing primarily with the behaviour of the primary coolant system and the emergency core-cooling system during loss-of-coolant accidents;

- fuel behaviour, dealing with how the fuel responds during various accidents;

- fission-product release and transport, dealing with the fraction of the products released from the fuel during an accident and how these migrate after they are released. Migration was strongly dependent on fission-product chemistry;

- fuel-channel behaviour, which examined how the pressure and calandria tubes behaved, especially during an accident;

- containment behaviour, dealing with how this system responded to the pressure rise during a severe LOCA, how well it retained fission products and how it coped with any hydrogen that might be produced.

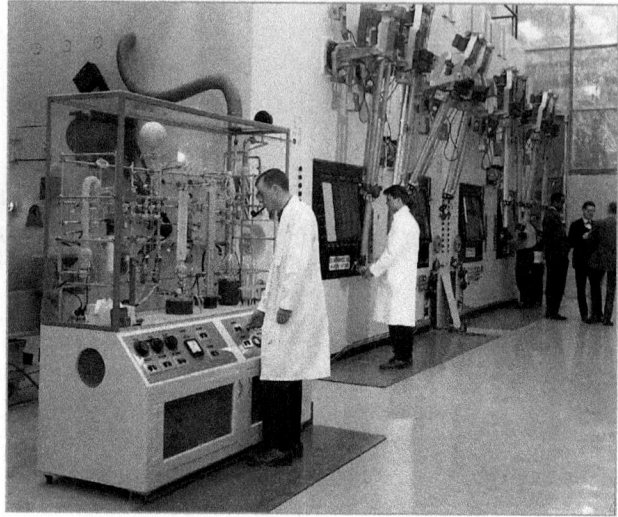

Roy Styles and Marv Ryz

Stanley Hatcher

Stanley Hatcher obtained his B.Sc. and M.Sc. in Chemical Engineering from Birmingham University. In 1954 he immigrated to Canada, obtained his Ph.D. in Chemical Engineering from the University of Toronto in 1958, and joined Atomic Energy of Canada Limited (AECL) as a research engineer at the Chalk River Nuclear Laboratories.

He spent 34 years with AECL, with 19 years at the WNRE. He became WNRE Site Head in 1978. In 1985, Dr. Hatcher became President of AECL Research, responsible for Canada's nuclear science and technology R&D programs. He became President and CEO of AECL in 1989 and for three years led the restructuring of the corporation towards its new emphasis in support of CANDU. Stan retired from AECL in 1992.

Stan was acclaimed internationally for his work in the nuclear field and in 2010 received the prestigious Global Award of the International Nuclear Societies Council. Stan was also a Fellow of the Canadian Academy of Engineering, the American Nuclear Society, the International Nuclear Engineering Academy, the Canadian Nuclear Society, the Chemical Institute of Canada and the Canadian Society for Chemical Engineering.

Reactor Corrosion and Activity Transport

Perhaps the most important issue related to safety in any reactor is the fraction of the fission products that will escape from the fuel during an accident, and where they will go. Thus, from the beginning, AECL investigated how fission products escaped from CANDU's uranium oxide fuel and how they migrate throughout the plant.

A program was started at Whiteshell in the mid-1970s. These studies concentrated on the tendency for iodine and cesium to form compounds and the stability of these compounds under LOCA conditions. The kinetics of the relevant thermodynamic relationships were derived and incorporated into computer codes.

These studies showed that under LOCA conditions in water-cooled reactors, iodine and cesium combine to form cesium iodide. Cesium also forms cesium hydroxide. Cesium iodide and cesium hydroxide are both salts that readily dissolve in water. Since there would be plenty of water present during a LOCA, only a small fraction of these fission products would be released. The results of this work explained the fission-product behaviour observed during the Three Mile Island accident.

The CANDU power plant has several large liquid systems: the primary cooling system, the moderator, the emergency coolant system and the dousing system. It is now believed that even in a worst-case accident the majority of the fission products that would escape from the fuel would be dissolved in water and that any release to the environment would be extremely small. In addition, all CANDU plants were designed to have highly reliable containment systems.

Front Row - Ara Mooradian, Howard Gilmour, Mike Wright, Roger Dutton, Roy MacFarlane, Frank Theriault; Middle Row - Bob Robertson, Frank Garrow, Brian Smith, Vic Serrata, Dennis Fitzsimmons; Back Row - Ed Olchowy, Dennis Penner, Bernie Gordon Jr., George Penner, Roy Styles, Des McCormac, Doug Benton, Bud Mager

Radioiodine Test Facility

Scientists at WL studied the behaviour of iodine under simulated conditions in the Radioiodine Test Facility (RTF). Their goal was twofold: to validate the assumptions made by Canadian nuclear utilities in predicting the consequences of worst-case nuclear accidents scenarios, and to develop accident-management strategies that the CANDU reactor was among the world's safest designs.

The RTF, built in 1987, included a 400-litre cylindrical reaction vessel with shielding for a cobalt-60 radiation source designed to reproduce the radiation fields that would be encountered during accident scenarios. The inner surface of the vessel could be changed by using liners made of stainless steel, carbon steel or concrete, to simulate different containment building surfaces and coatings. The temperature of the vessel could also be varied up to 80°C.

Tests were done using surfaces such as zinc primer, vinyl, polyurethane and epoxy. They showed that zinc primer provided an important absorbent surface for iodine. The presence of radiation also produced a large increase in the amount of iodine deposited on surfaces. Information from this program was used to improve the performance of CANDU containment buildings and provide more accurate assessments of the release of iodine from these buildings. Other key studies were to measure the behaviour of radioactive iodine under accident scenarios in light-water reactors, and to provide data on iodine behaviour to the European Communities' Phebus Fission Product Project. The RTF certainly made a valuable contribution to the world's knowledge on reactor safety.

Fuel Behaviour

The research programs were directed at obtaining a basic understanding of fuel behaviour under normal operating conditions and establishing limits for safe operation of the fuel.

Many experiments were done with operating parameters well beyond what were later specified as safe operating limits. Fuel was operated at power ratings high enough to produce central melting of the uranium oxide. Some defects occurred and in some cases the hot central part of the fuel came in direct contact with the coolant.

In a series of tests at Whiteshell, CANDU fuel bundles were heated in vacuum up to 1,600°C to study their mechanical behaviour at elevated temperatures. In these tests the bundles

Radioiodine Test Facility

slumped into contact with the pressure tube and bundle elements slumped into contact with their neighbours. However, there were no sheath failures and the bundle end-caps remained intact.

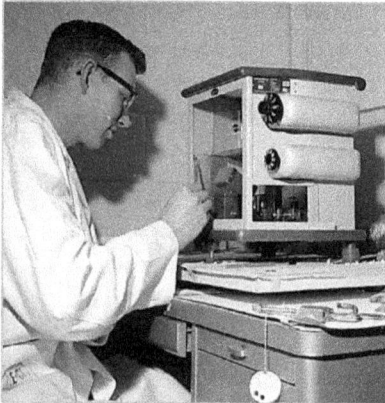

Jay Hawton

Gerry Walters

Advanced Fuels

In the early years it was believed the supply of economically recoverable uranium in the world was limited. Consequently, advanced fuel cycles would be a necessity. The first generation of power reactors were expected to be thermal neutron reactors fueled with either natural or slightly enriched uranium. The plutonium produced in those reactors would be separated and recycled to extend the available fuel supply. Early research programs included components on plutonium separation, fuel development, and the physics of specially designed reactors.

In Canada, early work on advanced fuel cycles was focused on thorium breeder fuel cycles. As early as 1950, W.B. Lewis was writing papers in which he described low-cost electricity for thousands of years through the use of uranium in fast reactors and thorium in heavy water reactors.

Fuel Channel Performance - Nitheanandan
Thambiayah, Brock Sanderson, and Roger Dutton

Immobilized Fuel Test Facility

Ralph Green

Ralph Green (B.Sc., M.Sc. (Dalhousie), Ph.D. (McGill)) joined AECL at the Chalk River Laboratories in 1956, working first in reactor physics at Canada's first research reactor ZEEP, then in accelerator physics, and later as head of the Reactor Control branch.

In 1979 he transferred to AECL's head office in Ottawa as a senior advisor. In 1982 Ralph was appointed vice-president and general manager of the Whiteshell site. In 1986 he was appointed vice-president of Reactor Development, responsible for all reactor-related R&D in AECL.

After retiring in 1991, he contributed to the 1997 book on the history of AECL, "Canada Enters the Nuclear Age". Ralph was a charter member of the Canadian Nuclear Society, having joined in 1980. Since his retirement, Ralph has been active in the Ottawa Branch of CNS.

As the demand for uranium increased, more was found, and by the mid-1960s it was clear that enough uranium was available to fuel the world's reactors for many decades on once-through cycles. Advanced fuel cycle work now had to be justified on economic grounds rather than resource grounds. Countries with adequate uranium reserves phased out advanced fuel work. In countries with poor uranium reserves, work continued but the emphasis shifted to recycling in thermal reactors. Fast-reactor programs continued in some countries, most notably France and Japan. In Canada, there was an early flurry of activity on plutonium and uranium-233 separation and on recycling plutonium in the NRX reactor. The work was confined to the laboratory scale, with the objective of verifying the capability of the CANDU reactors to economically use advanced fuel cycles when that became necessary.

In 1968, a government/industry study concluded that plutonium recycle could be marginally attractive by 1975. Again in 1976, when the pressure to work on disposal of used nuclear fuel was mounting, AECL asserted to the government that it would be irresponsible to dispose of the fuel without recovery of the plutonium, and that the recycle program should be expanded. These studies increased the profile of fuel reprocessing and the issues of disposal of reprocessing waste.

W. B. Lewis became a champion of the CANDU-OCR concept in the late 60s, which he felt was the best reactor for his Valubreeder fuel cycle, based on the recycling of uranium-233 and thorium topped up with plutonium from uranium-fueled reactors. Since organic reactor technology was primarily a Whiteshell responsibility, Whiteshell also undertook to verify the technical aspects of the Valubreeder fuel cycle.

In the 1970s, twenty-four bundles containing thorium and uranium-235 were irradiated in WR-l. These bundles performed well and reached burn-ups of 900 and 450 MWh/kg U respectively. Unfortunately, because of the characteristics of WR-1, the fuel could be operated only at modest power ratings.

By the mid-1980s the main interest in CANDU advanced fuel cycles was from foreign owners and potential foreign customers. AECL was also searching for a way to demonstrate the technology on a large scale. To do this, they turned to the slightly enriched uranium (SEU) fuel cycle, which provided a 20% savings in fueling costs over the natural-uranium fuel cycle in Canadian reactors. The SEU cycle demonstrated the fuel behaviour and fuel management aspects of advanced fuel cycles on a large scale. These experiments were done in NPD and WR-1.

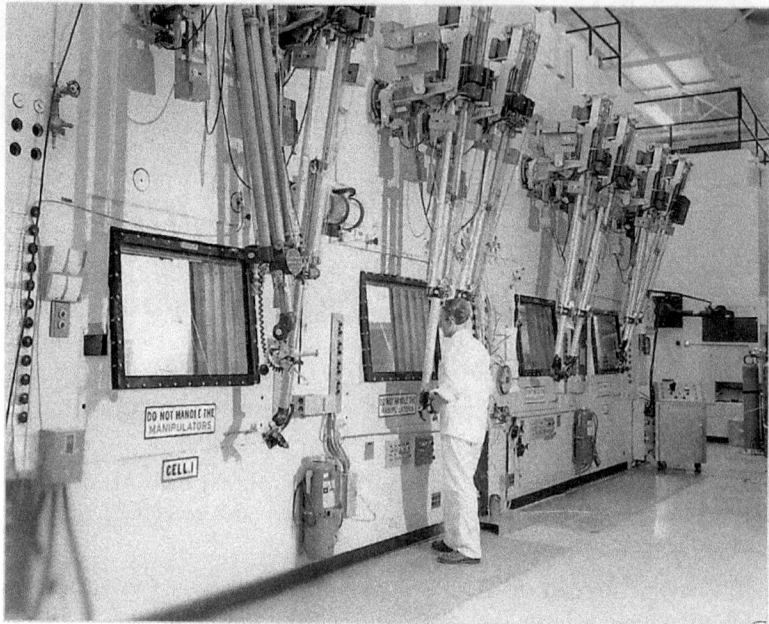

Bernie Komadowski - Safe handling of radioactive samples

The excellent behaviour of natural uranium taken to high burn-up and the behaviour of LEU and thorium/ uranium-235 led to a high degree of confidence that fuel of the conventional CANDU design would give good performance in any of the advanced fuel cycles envisaged.

Maple Research Reactors

In the mid-1980s's, a group of WL reactor physicists, supported by thermalhydraulics analysts, began to develop a new concept in research reactors. There were several motivations behind this initiative. Worldwide, the first generation of small, research reactors were beginning to show their age. Secondly, the USA was starting an initiative of their own - the RERTR, or Reduced Enrichment for Research and Test Reactors, to convert the existing research reactors from highly-enriched (nominally 93 wt% ^{235}U in total uranium) fuel, into low-enriched (less than 20 wt%) uranium fuel. The aim of the American program was to further nuclear weapon non-proliferation. Another goal of the AECL program was to develop a research reactor to market to developing countries as an ideal way to develop national reactor expertise, in preparation for an eventual foray into the acquisition of a power reactor(s). A prototype of the new reactor would be built at the Chalk River Laboratories to take over the production of medical radio-isotopes from the aging NRX and NRU reactors.

Initially, there were two variants of the new research reactor. The first was the MAPLE (Multipurpose Applied Physics Lattice Experimental) concept, with an anticipated thermal power rating of 10 to 25 MW, and a smaller, lower-cost version, SUGAR (SLOWPOKE Uprated for General Applied Research) MAPLE, with an anticipated thermal power rating of approximately 1 to 10 MW. The two-model MAPLE concept quickly merged into the single, MAPLE concept, which retained the 10 to 25 MW design power.

The basic MAPLE concept consisted of an open-tank-type reactor assembly within a light water pool. The MAPLE reactor had a heavy water (D_2O) reflector tank and hexagonal fuel bundles, containing low-enriched 19.75 wt% uranium in a uranium-silicide-aluminum matrix. The reflector tank would contain multiple material irradiation sites and could be penetrated with the snouts of neutron beam tubes for external neutron scattering experiments (as NRU featured). The reactor core would be designed for a variable number of driver fuel bundle sites, to support the range of design power outputs intended.

In 1985, the Korean Atomic Energy Research Institute (KAERI) decided to purchase a 30 MW MAPLE reactor, called HANARO. KAERI sent a team of approximately 15 reactor physicists and thermalhydraulics analysts to WL for 18 months to work with their WL counterparts on the design of HANARO. HANARO first went critical on February 8, 1995.

Meanwhile, the WL staff were starting the design of MAPLE-X10 (MX-10), a 10 MW reactor to be designed for radioisotope production at CRL. In fact, there were to be two MX10 reactors, one for primary radioisotope production (primarily Molybdenum-99, or Moly-99) and the other as a standby, to ensure the reliable production of isotopes for medical purposes.

Understanding the Physics – George Penner, Heidi McIlwain, Ian Gauld, Bruce Wilkin, Herb Rossinger, Shannon Worona and Bob McCamis

In parallel, the MAPLE team designed slightly modified versions of the basic reactor, in response to invitations to build on research reactors in several countries, including Egypt, Australia, Thailand, and Indonesia. Variations included different thermal power output, different locations (neutron fluxes), sizes and purposes of in-reflector irradiation sites, and properties of neutron scattering beam tubes. One possible neutron beam tube featured a cryogenic insert to produce 'cold' (i.e., sub-thermal) neutrons for various experiments. In the end, unfortunately, AECL did not win the bid for any of these research reactors, aside from the initial KAERI MAPLE.

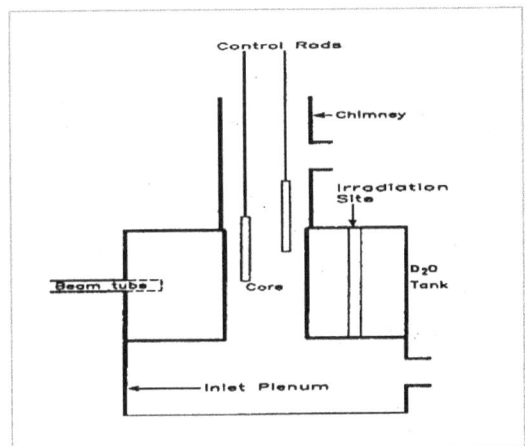

The MAPLE-X10 reactors were constructed at CRL, with first criticality occurring on February 19, 2000. However, a problem soon developed when it was

Vertical view of a generic MAPLE reactor.

determined that the MX10 Power Coefficient of Reactivity (PCR) was slightly positive, in contrast to a slightly negative prediction. The PCR is defined as the change in reactivity of the fuel as the reactor power increases; in other words, if the reactor power increased a small amount, a positive PCR meant that the reactor power would want to increase even further. The MX10 was designed with negative coefficient, including the PCR, which would have meant that the reactor would be self-regulating, with the reactor wanting to decrease power following a small power increase. The magnitude of the PCR was very small, meaning that the MX10 control system could easily be designed to control the reactor.

The MX10 commissioning tests were temporarily halted while detailed computer simulations and bench-scale tests tried to find design modifications to produce a negative PCR. Eventually, it was concluded that the positive PCR was caused by the small size of the MX10 core. Differential temperatures between the inside and outside of the individual fuel elements caused bowing of the fuel elements, which led to the positive PCR. However, by this time, regulatory opinions had hardened, and the projects were cancelled by the Canadian government before redesigned fuel bundles could be tested.

Thermal Hydraulics

One of the most important safety-related issues concerns the rate of coolant loss from the primary circuit in the event of a major pipe break. A unique feature of the CANDU design is

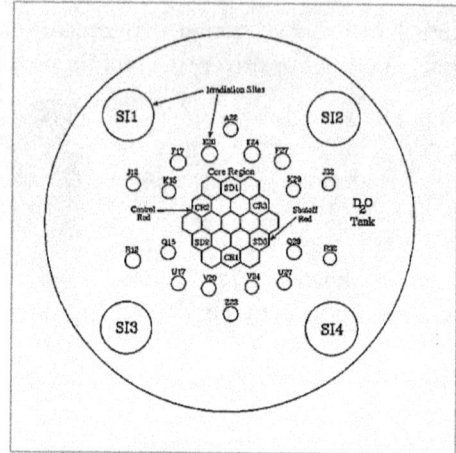

Schematic diagram of MAPLE-X10 reactor.

the flow-distribution headers. These headers are long, large diameter pipes that connect the inlet feeder pipes to the coolant pumps (called the inlet headers), or the outlet feeder pipes to the steam generators (the outlet headers). In a LOCA, emergency coolant is assumed to be injected into these horizontal headers. Flow stratification in the headers could affect the distribution to the feeders during this emergency coolant injection. To investigate this effect, a full-scale header facility representative of the Pickering plant was constructed at Whiteshell in 1974. In this facility, the header temperature could be brought to the desired value using electrical heaters. During coolant injection, the liquid level and pressure were measured at several positions along the header, and feeder flow rates were monitored for a range of conditions.

In the late 1970s, a small-scale header facility was constructed at Whiteshell. A key feature of this facility was the use of two transparent windows that permitted direct observation of phenomena that could occur in the two-phase (steam/water) mixture during emergency coolant injection. To further explore complicated phenomena such as "vapour pull through", two header facilities were also constructed, one at AECL's engineering laboratories at Sheridan Park and the other at Whiteshell, using Lucite for the headers. Coolant blowdown experiments were done to study the effect of pipe-break size on depressurization rate, flow distribution and heated-section temperature. Experiments were also done with cold water injected simultaneously into each header at the end of depressurization.

John Findlay running reactor loop experiments

All of these separate-effects tests greatly improved understanding of what might occur during a LOCA. However, to check the complete LOCA analysis code, integrated tests were required. The first such test was performed in the RD-4 facility at Whiteshell, commissioned in 1974. This was a small-scale recirculating water loop containing two pumps, two tubular heated sections to simulate fuel channels, and two heat exchangers.

The RD-4 facility was succeeded in 1977 by the much larger RD-12 facility, which had heated sections 4-m long with bundles of seven electrically heated elements to simulate the fuel. This facility also had recirculating U-tube type boilers, with an interacting secondary circuit and a much wider temperature range of operation than RD-4. A pressurized cold-water injection system was also provided, to supply water simultaneously to the four headers when the loop pressure fell below a set value, thereby simulating a LOCA with emergency coolant injection.

The experimental program included tests with various pipe-break sizes, located at different points in the loop circuit, various cold-water injection pressures, and several modes of boiler cooling. In all of these experiments detailed measurements were made to determine coolant flow rate, pressure, temperature

Model of RD-4 and RD-5 Loops

and density distribution throughout the loop, as well as fuel-element surface temperatures, and differential pressures across various loop components.

RD-14 and RD-14M Reactor Loop

The RD-12 facility was followed in 1983 by an even larger facility, called RD-14, which was a model of a primary coolant loop with the various components arranged to reproduce the gravitational effects in a CANDU plant. The large size and height of RD-14 required the construction of a special building to house it at Whiteshell. Its operation was quite spectacular, since each blowdown test resulted in the release of a large quantity of steam through a pipe on the roof of the building.

RD-14 consisted of two full-scale, full-power fuel channels, each containing a full-length, 37-element, electrically-heated bundle to simulate the fuel, plus full-sized feeders and two full-height steam generators, all arranged in the CANDU figure-eight configuration. The steam generators had full-sized U-tubes, but the number of tubes was reduced in proportion to the number of heated channels, to give the correct heat transfer area per fuel channel. The loop was designed so that fluid mass-flow rate, transit times and pressure/enthalpy terms in the primary system of the loop were the same as those in a typical CANDU under both forced- and natural-circulation conditions.

The RD-14M Facility (upgraded RD-14) was operational in 1987. The RD-14m facility was one of the largest of its kind in the world, with an overall height of about 34 m. It was used to investigate many LOCA related phenomena, such as:

- blowdown tests for many different pipe break sizes, where the pressure difference across the reactor core becomes close to zero, resulting in a stagnant flow condition in the fuel channels for an extended period (tens of seconds);

- two-phase thermo syphoning tests, to study the situation that would arise from a loss of the coolant pumps, or during small-break LOCAs, where the fuel channels are cooled mainly by natural circulation.

Thermal Hydraulics Building

The RD-14M Facility was flexible; it could be set up to simulate a CANDU 6® reactor, the Enhanced CANDU 6, the Darlington reactor, and the ACR-1000 reactor. The facility was designed to produce similar fluid mass flux, transit time, pressure, and enthalpy distributions in the primary system as those in a typical CANDU reactor, under both forced and natural circulation conditions. This facility was also used to perform several secondary-side depressurization tests.

Many useful results were obtained from the tests done with the RD-14 and RD-14M facilities and these were used to refine and verify computer models, such as the CATHENA code. The results of these tests showed that the fuel sheath temperature would not exceed about 600°C for all LOCAs. Hence fuel should not fail, as long as the emergency coolant system was available.

Improving Reactor Designs

The CANDU safety program evolved over the years to address the many

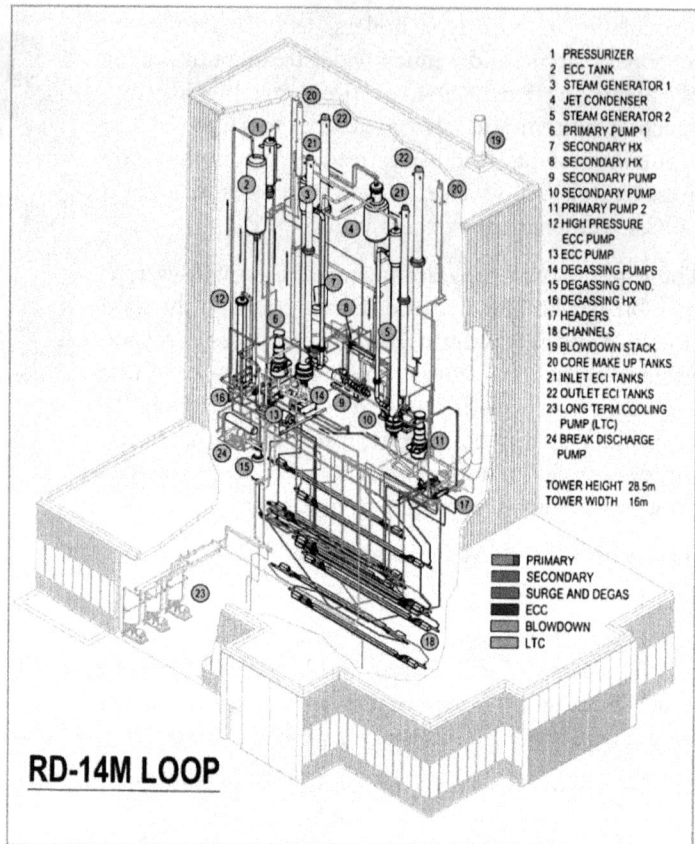

RD-14M LOOP

1 PRESSURIZER
2 ECC TANK
3 STEAM GENERATOR 1
4 JET CONDENSER
5 STEAM GENERATOR 2
6 PRIMARY PUMP 1
7 SECONDARY HX
8 SECONDARY HX
9 SECONDARY PUMP
10 SECONDARY PUMP
11 PRIMARY PUMP 2
12 HIGH PRESSURE ECC PUMP
13 ECC PUMP
14 DEGASSING PUMPS
15 DEGASSING COND.
16 DEGASSING HX
17 HEADERS
18 CHANNELS
19 BLOWDOWN STACK
20 CORE MAKE UP TANKS
21 INLET ECI TANKS
22 OUTLET ECI TANKS
23 LONG TERM COOLING PUMP (LTC)
24 BREAK DISCHARGE PUMP

TOWER HEIGHT 28.5m
TOWER WIDTH 16m

PRIMARY
SECONDARY
SURGE AND DEGAS
ECC
BLOWDOWN
LTC

questions that were raised by designers and regulators as the various CANDU plant designs evolved and proceeded through the licensing process. The safety programs confirmed the soundness of the original CANDU concept. The results obtained from these programs increased our confidence in the safety of the CANDU reactor and showed that even for low-probability accidents the regulatory standards could be met.

Hydrogen

Hydrogen generation and behaviour is an important safety issue in nuclear power plants. Rapid combustion of hydrogen can produce high temperatures and high pressures in confined spaces. Research on hydrogen was performed for many years at Whiteshell.

There are two main sources of hydrogen to consider in a CANDU power plant. The exothermic reaction between the Zircaloy fuel sheathing and steam might occur during a combined LOCA/ Loss of Emergency Coolant (LOEC) accident to produce hydrogen. The radiolysis of heavy-water coolant and moderator may also lead to a buildup of hydrogen and oxygen in the helium cover gas in the calandria vessel.

AECL scientists and engineers were cognizant of the potential hazard posed by hydrogen combustion, and a special facility to study hydrogen behaviour in containment, called the Containment Test Facility (CTF), was built at Whiteshell in 1980. The CTF comprised two large-scale vessels, a 2.3-metre diameter sphere and a 5.3-metre-long, 1.5-metre-diameter cylinder that simulated containment structures. They both had a pressure rating of 1.0 MPa. These vessels were used for large-scale experiments to determine pressure transients and integrated combustion effects. The facility also included a 28-cm diameter, 9-m long combustion pipe that could operate at 10 MPa. Obstacles could be mounted inside this pipe to induce flame acceleration.

The CTF was used to determine the conditions under which hydrogen combustion could occur within containment, and to acquire a database to validate the computer codes used to predict the peak pressures that might be produced as a result of hydrogen combustion during an accident and to assess the integrity of the containment building and the calandria vessel.

The experimental program investigated all aspects of hydrogen combustion, including such factors as flammability limits, laminar burning velocities and the effects of gas diluents, including steam, on these. Also studied were flame acceleration due to venting, the effectiveness of igniters that might be used to produce controlled burning of hydrogen, and the effect of turbulence on flame propagation.

Containment Test Facility

More fundamental studies of the detailed physics and chemistry of hydrogen combustion were also performed. Studies investigated turbulent burning velocities, detonation limits and the transition from deflagration to detonation. While a full-scale detonation is not possible in a CANDU containment building during a LOCA, the possibility of local detonations was a concern. In localized areas of high hydrogen concentration, detonation may be externally induced (e.g., by a local high-energy discharge) or self-induced by the transition of a deflagration to a detonation. These experiments led to the derivation of mechanisms and criteria for the transition to detonation in the presence of typical obstacles.

All of these studies have shown that it is extremely unlikely that conditions could arise where hydrogen combustion could threaten the integrity of a CANDU containment system, if proper precautions are taken.

The Large-Scale Vented Combustion Test Facility (LSVCTF) was a 10-m long, 4-m wide, 3-m high rectangular enclosure. The test chamber, including the end walls, was electrically trace-heated and heavily insulated to maintain temperatures in excess of 100ºC for extended periods of time. The combustion chamber could be subdivided into multiple compartments. The LSVCTF was designed to quantify effects of key thermodynamic and geometric parameters affecting flame propagation and pressure development during vented combustion. The facility was also used to test and qualify PARs. Many of the features of the LSVCTF made it a one-of-a-kind facility in Canada. Some of these features included a variable

Large Scale Vented Combustion Test Facility

vent opening, removable end walls, accurate control of initial thermodynamic conditions and variable geometric configurations between single or inter-connected multi-rooms, which mimicked rooms in nuclear reactor containment.

Ceramic Reactor Project and Hydride Cracking

In 1968 a small group of materials scientists were hired to evaluate the concept of a ceramic pressure tube, the idea being that if it could be made ductile and retain its high temperature strength, improvements in thermal efficiency of the reactors would be attained. Projects to evaluate the mechanical properties of ceramics and to look at strengthening mechanisms such as microstructure control, fibre re-enforcement and mechanisms of sub-critical crack growth (SCCG) were developed. While staff was somewhat sceptical of developing and licensing a ceramic pressure tube, there was a lot of interest in high temperature applications for other industries. Among other things was the development of useful techniques for determining SCCG behavior and also a theory for diffusional crack growth.

In 1973 a pressure tube had burst in the Pickering #1 Reactor. The rupture was quickly found to be caused by migration of hydrogen to a cold spot forming hydride. The pressure tube had sagged into contact with the cold calandria tube. A hydride crack was initiated that slowly grew to critical size and the tube ruptured. All materials scientists were instructed to join a fuel channel working party and address the issue at once. It was worried at the time that the entire CANDU concept could be in question.

The failure mechanism was called delayed hydride cracking (DHC). Delayed hydride cracking is a sub-critical crack growth mechanism occurring in zirconium alloys. Hydrogen in solution in the zirconium alloy is transported to the crack tip by diffusion where it precipitates as a hydride phase. When the precipitate attains a critical condition, related to its size and the applied stress, a fracture ensues and the crack extends through the brittle hydride and arrests in the matrix. Each step of crack propagation results in crack extension by a distance approximately the length of the hydride.

The decision was made to re-tube the first two Pickering reactors with Zr 2.5%Nb, the alloy used in all subsequent reactors. The newer reactors seemed fine and were kept running as normal.

However, hydride cracks were soon discovered near the end fittings of the Zr2.5%Nb reactors due to errors in fabricating the rolled joints connecting the pressure tubes to the end fitting. One tube actually burst while at pressurized shutdown. Thus not only temperature gradients but high stresses also had the capability to attract hydrogen to regions where it would exceed the hydrogen solubility and form hydrides. Fortunately for AECL, when pressure tubes started failing and developing cracks, our group was able to immediately start using our techniques to characterize SCCG in zirconium alloys and

Bob Shewfelt measuring material properties

develop a physical theory for DHC crack velocity, which reflected the experimental measurements we made in our labs. CANDU's kept their licenses as new operating procedures were established to minimize hydrogen pickup and ensure that hydrogen solubility was not exceeded during operation.

Whiteshell became a world centre for DHC studies in the 1980s and 90s, both in zirconium alloys and ceramics (M.P. Puls, Eng. Mat. Series, Springer-Verlag, 2012). Key milestones and conclusions from this work included:

- A modified theory for sub-critical crack growth in ceramics was developed to account for hydrogen diffusion to high stress areas.
- Hydrogen preferentially diffuses to a region of high stress.
- There is no upper temperature limit for hydride cracking as long as hydrogen is present.
- Crack velocities were very high (mm/day) at 300°C.
- The diffusion model accurately estimated the crack velocities.
- Guidelines for reactor operations were established to reduce the risk of hydride cracking.

Whiteshell staff worked closely with our colleagues at Chalk River, Sheridan Park and Ontario Hydro Research. Several working parties were formed under the COG (CANDU Owners Group) Fuel Channel Technical Committee. This multi-site cooperation eventually led to amalgamation of branches across sites and with branch managers responsible for employees at both sites. This inter-site co-operation worked successfully in the Fuel Channel Division and was later duplicated by Reactor Safety Research Division with headquarters at Whiteshell but also including the Fuels and Materials Branch at Chalk River.

Pressure Tube Elongation

Changes in the shape of pressurized tubes caused by operating temperatures and pressures are enhanced by fast neutron irradiation. Pressure tubes in CANDU reactors and WR-1, were monitored periodically over 20 years in the 1970s and 80s. The deformation occurring during steady-state irradiation creep and growth was modelled, taking into account the presence of intergranular stresses. The predictions from a deformation equation based on data from the Pickering and Point Lepreau Nuclear Generating Stations and the WR-l, Osiris, DIDO, and NRU test reactors were in good agreement with actual measurements. This equation has been employed as a material subroutine in the 3-D finite element code used for predicting the detailed shape change of pressure tubes in CANDU reactors.

Managing Nuclear Waste

When Whiteshell started up in 1961, it was not blessed with sandy uplands well above the water table. Its property consisted of about 35 km^2 of low-lying forest and cleared farmland near the north-east boundary of the plains area of Manitoba, on the east bank of the Winnipeg River. The regional overburden was clay, overlying a compact layer of glacial till and bedrock. The water table was usually within 2 m of the surface, so it was not possible to find an area where waste could be buried above the water table. However, on the property, about 4 km from both the laboratory buildings and the Winnipeg River, was a hydrogeological discharge area; i.e., an area where the groundwater flow was to the surface. This area was chosen for the waste management area (WMA) because as water flowed to the surface it was channelled into a discharge ditch and monitored to ensure that excessive levels of radionuclides were not released to the Winnipeg River. An area of 4.6 hectares was fenced off for this purpose and no other area has subsequently been required.

Trenches for low-level wastes, concrete bunkers and concrete stand-pipes for medium-level wastes, and stainless-steel tanks for high-level waste liquids were standard options for handling the different categories of waste. To minimize weather exposure and problems related to the high water table, low-level wastes were buried in the spring and in the fall. In the interim periods they were held in a steel building.

The WMA was also equipped with an incinerator to burn organic liquids, primarily organic coolant used in the WR-l reactor. The incinerator used a vortex burner, which allowed incineration without visible smoke or ash. Radioactive emissions were controlled by limiting the radioactivity in the feed to less than 15 Bq/ml. Organic liquids with radioactivity above that level were stored in stainless-steel drums on a concrete pad until the radioactivity decayed to the required level.

The waste handling facilities were modified over time to reflect increased knowledge of the research and the area. Concrete bunkers and stand-pipes were surrounded by bentonite clay to keep water out and adsorb any radionuclides that might be released. Concrete stand-pipes were lined with galvanized steel to further protect against water ingress. Waste from irradiated fuel was stored in concrete canisters rather than concrete stand-pipes.

The newest facility in the WMA is the Shielded Modular Above-Ground Storage (SMAGS) Building. The first SMAGS was constructed in 2010, providing 4000 m³ of low-level radioactive waste storage.

Terry Rummery

Terry Rummery's academic achievements began with an Honours B.Sc. in Engineering Chemistry, Queen's University, 1961, a Ph.D. in Physical Chemistry, Queen's University, 1966, and a National Research Council Overseas Post-Doctoral Fellowship, University College, London, UK, 1967.

Terry joined AECL in the early 1970s, working on a number of chemistry related research programs. He progressed to Research Chemistry Branch Manager, Whiteshell, the Waste Management Division Head, AECL VP and President of Atomic Energy Canada (Research). Terry led the program to develop the safe disposal of used nuclear fuel. For his work he was awarded an Honorary Doctor of Science degree from Queen's University in 1993. In 1994 he received the W.B. Lewis Medal for contributions to nuclear science and engineering. His other accomplishments include Chairman of the Board for the Chemical Institute of Canada, 1998, and Fellowships in the Canadian Academy of Engineering and the Canadian Nuclear Society.

The Active Liquid Waste Treatment Centre (ALWTC; Building #200), began operation in 1963, receiving low-level liquid waste effluent from operating facilities (WR-1, Shielded Facilities, B300 Research Laboratories, Laundry/Decontamination). The liquid effluents were transferred via underground piping to the ALWTC. The ALWTC included a medium-level liquid waste processing system which concentrated the waste stream. The resulting concentrate was solidified and stored at the WMA.

Used Fuel Storage

The WR-1 reactor had a very small used-fuel bay. By about 1972 the bay was filling up. Whiteshell had to either build a new bay or devise another method of storing fuel. They decided to experiment with dry storage in reinforced concrete canisters.

The attraction of the canisters was that they could be built as needed and thus would avoid the large up-front expenditure that would be required for a new bay. The concern was that the thermal gradient across the concrete (hot on the inside, cold on the outside) would cause the inner concrete to expand and crack the outer concrete causing the exterior surface to weather rapidly or lose shielding because of a crack through the wall. Fine exterior cracks were expected to relieve stresses, but it was the size and behaviour of the cracks that was important. The objective of the development program was to demonstrate that used fuel could be stored in this way and verify prediction methods to improve future designs. Four experimental canisters were built; two cylindrical and two square.

Dry Storage Canister

Although it was clear that cylinders would be the most efficient in terms of heat transfer and utilization of material, all power-reactor fuel-handling equipment was based on a rectangular configuration. One canister of each design was to hold fuel and the other was to be equipped with electric heaters to test the design limits of the concept.

The canisters were 5.3 m high with interior mild steel liners. Fuel was placed in steel baskets and a lid was welded on. The baskets were filled with inert gas to prevent oxidation of the fuel. Six baskets were placed in each canister. The design heat load was 2 kW, about 4.4 Mg of five-year cooled fuel.

Tests on the electrically heated canisters showed that the designs were practical and conservative. Cycling tests to represent summer and winter extremes showed that cracks did not penetrate deeply and would close up as the heat decayed. Canisters were then filled with fuel: the cylindrical one with WR-1 fuel and the square one with Douglas Point fuel. These tests confirmed the viability of the concept and it was adopted for the storage of WR-1 fuel. This technology has since been used for the storage of Gentilly-l, Douglas Point and NPD fuel when those reactors were shut down.

Loading Demonstration Canister

Research Chemistry

The history of research chemistry at Whiteshell from its origins in 1961 as a group within the Whiteshell Division at Chalk River, through creation of a Research Chemistry Branch in 1970 up to the branch's final form in the 1990s, has been described by N. H. Sagert (AECL Report #11027). While programs changed greatly over 30+ years, there was a common thread of mission-oriented fundamental chemical research in a small number of key chemical disciplines. From the arrival of a new electron accelerator at Whiteshell in 1963 until the mid-1970s, radiation chemistry was the dominant topic within the branch and its predecessor groups. The unique achievements in this field form a large portion of this account. Later research programs were based on the need for expertise according to the states of matter: solid, liquid, and gas phases, as well as interfacial (especially solid/liquid) chemistry, often complicated by the need to obtain accurate measurements at high temperatures and/or pressures. Strong ties were developed with related research groups in other countries, especially France, Sweden, the U.K., and the U.S.A., and with international organizations such as the Nuclear Energy Agency of the Organisation for Economic Co-operation and Development.

The primary objective of the Radiation Chemistry group at Whiteshell was to improve AECL's capability to predict and assess the effects of ionizing radiation on liquids, gases, and solids of interest to Canada's nuclear power plants. The program initially focused on understanding the effects of radiation on

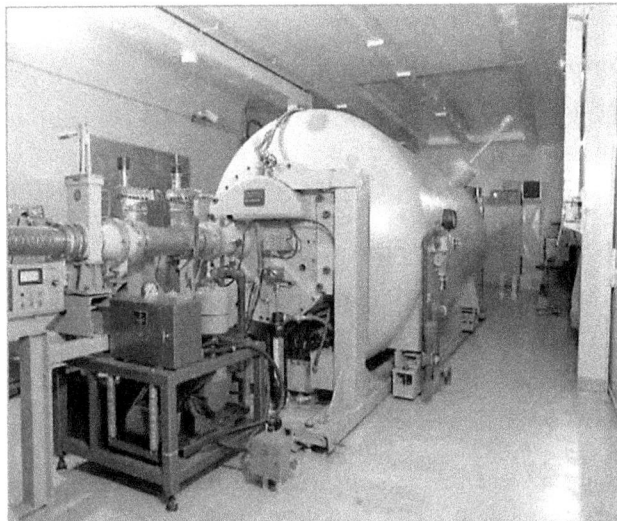
Van de Graaff Accelerator

heavy water (D_2O) and organic coolants. Later, Whiteshell's program in long-term nuclear fuel waste management led to an interest in water radiolysis because of the potential effects of irradiated groundwater on used fuel container surfaces, and on fuel oxidation and dissolution, in an underground repository.

Radiation chemistry also proved to be a key to understanding the behaviour of iodine from a reactor safety viewpoint, especially the radiation-induced formation of volatile organic iodides in containment.

Electron Spin Resonance and Pulse Radiolysis Chemistry

The first of the instrumental advances for research on radiation chemistry was the electron spin-resonance (ESR) technique introduced at Chalk River around 1963. The ESR technique was introduced at Whiteshell in 1967.

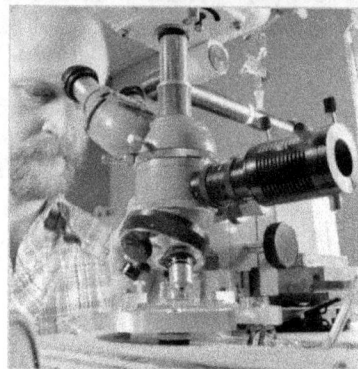

Pulse radiolysis was introduced at Chalk River in 1969 and at Whiteshell in 1972, using electrons from Van de Graaff accelerators. At Whiteshell, the focus was mainly on the chemistry of the free radicals and solvated electrons produced in radiolysis and a wide variety of compounds were studied. Of particular note were the direct observations of several of these species using a combination of ESR and pulse-radiolysis techniques. The fundamental aim in all these pulse radiolysis experiments was to test a theoretical prediction of the effects of changes in magnetic field strength on the light emitted by the molecular fragments in liquids during their brief existence; less than a microsecond following irradiation pulses.

Mike Quinn

Radiolysis of Terphenyls

While the WR-1 reactor used heavy water as moderator, it used an organic liquid, Monsanto's HB-40, as coolant. A possible negative factor to using an organic coolant was the production of organic degradation products which could affect coolant flow, activity transport, and corrosion. Early radiation chemistry studies focused on understanding the breakdown products of HB-40. Decomposition of the individual terphenyls was measured under both thermal and radiation conditions using direct beams from the 1.5 MeV Van de Graaff accelerator. The thermal and radiation effects were found not to be additive. The thermal decomposition was found to be enhanced by prior irradiation, while the radiolytic process underwent distinctive changes at elevated temperatures.

Gamma Ray Energy Absorption

Relative rates of gamma ray absorption in different materials were computed for a variety of gamma-ray spectra for different reactors. The program led to estimates of the mean gamma-ray mass energy absorption coefficients for different materials and gamma ray energy spectra.

Radiolysis of Water Vapour

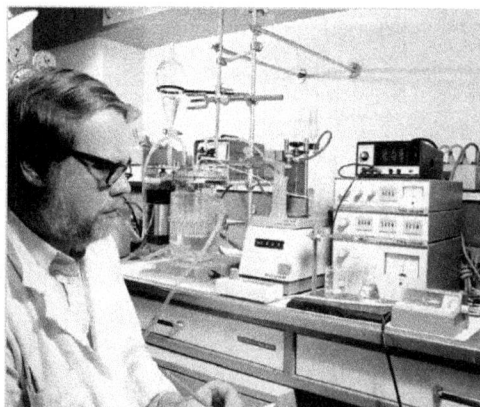

A program was designed to investigate yields of intermediates products in irradiated water vapour. The discovery of the solvated electron (known as the hydrated electron in water) as the precursor of the H atom in irradiated solutions had opened up a new field of study in aqueous solutions. However few studies had been carried out in the vapour phase, where hydrated electrons were also expected to be found. The program was designed to

Norm Sagert

measure ion, radical and molecular yields in irradiated water vapour, to correlate the electron yields with known constants such as reaction cross-sections, and to identify differences between the effects in the vapour and liquid.

Initially water containing electron scavengers was irradiated at 125°C using gamma or X-rays. This enabled the yield of H atoms arising from solvated electrons to be estimated. Water vapour was then irradiated in the presence of methanol and various electron scavengers, which led to estimates for the yields of H atoms arising from solvated electrons, H atoms from processes not involving solvated electrons, and molecular H_2, which was unaffected by the addition of scavengers.

Dave Torgerson

Later the studies were extended to mixtures of water and 2-propanol, in the presence of various additives, to identify general properties and reactions following irradiation of molecules containing the OH group. These studies confirmed the results previously obtained in water vapour, and led to estimates of the yields of solvated electrons, H and H_2 in pure 2-propanol and its mixtures with water. An extension of these studies in the presence of N_2O showed that a reduction in solvated electrons did not equate with the production of N_2 at high alcohol concentrations, i.e. the N_2 yield cannot always be taken as a measure of solvated electrons in water-polar organic mixtures.

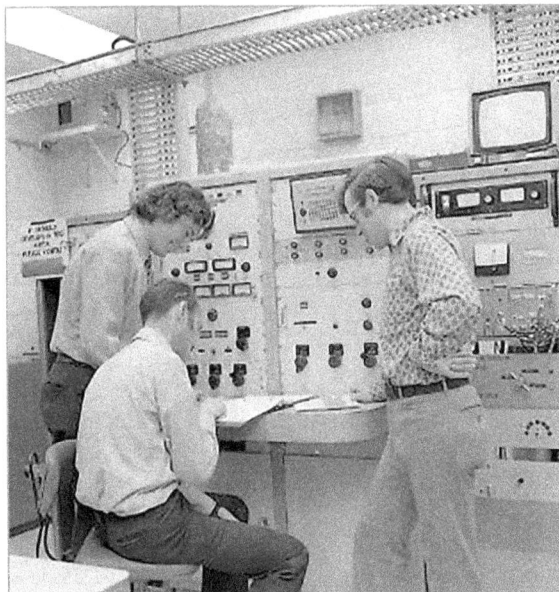

Bob Dixon, Vince Lopata and Jack Frederick

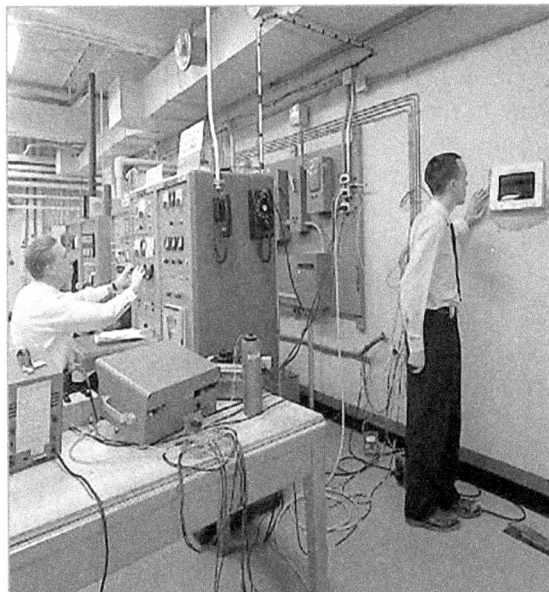

Mike Tomlinson and Ron Atherley

Computer Modelling of Water Radiolysis

A program was established to estimate the concentrations of intermediates and products in irradiated water through computer simulation and compare the results with existing experimental data. The objective was two-fold:

- to evaluate the use of computer modelling in assessing the effects of water radiolysis products in the reactor heat transport system and the moderator system, and

- to help assess the effects of irradiated groundwater on used fuel container surfaces in an underground waste repository.

Initially, computer programs developed in the U.S. and at CRL were compared using literature data for yields and reaction rate constants. Comparisons showed that the radiolysis of pure water and neutral solutions containing added H_2, O_2 and/or H_2O_2 at room temperature could be modelled satisfactorily using published data for primary yields and reaction rate constants.

Fast Reaction Kinetics and Mechanisms

Adding a 5-μs pulse capability to the 1.5 MeV Van de Graaff permitted the study of decay kinetics and mechanisms in liquids. However, the primary reducing species, the solvated electron, decayed at a rate much faster than this pulse width. The purchase of the 4 MeV Van de Graaff accelerator added a 5-ns pulse capability. Using optical absorption techniques, it was now possible to measure primary yields and decay mechanisms of solvated electrons in liquids. At the time, it was one of only three such systems operating worldwide.

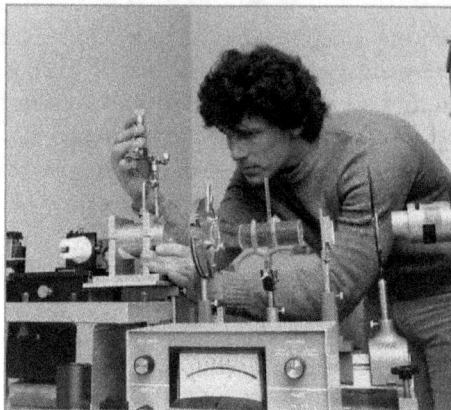

Grant Koroll – Physics at work

The work at Whiteshell focused on the production, yield, decay and reaction mechanisms of the solvated electron in H_2O, D_2O, simple alcohols and other organic liquids. By adding a variable temperature control, properties could be measured from 150° to 450°K. One of the most useful properties was the absorption spectrum of the solvated electron, which was measured in heavy water over the range 292-445°K. Knowing its absorption spectrum, the production, yield and decay kinetics of solvated electrons could be studied at different temperatures. Unfortunately, the early reaction processes were extremely fast and at higher temperatures could be complete by the end of the pulse. Picosecond pulse radiolysis would overcome this drawback but it was at an early stage of development at the time.

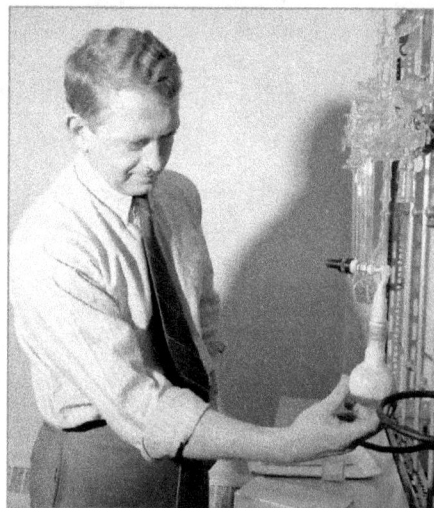

Graeme Strathdee

To obviate the difficulty, the decay kinetics and yields were studied at lower temperatures, to slow down some of the early reactions. Slowing down reactions would help us differentiate among free solvated electrons, solvated electrons that recombine with positive ions, and those electrons that recombine before they have time to become solvated.

Initially, a study of pulse-irradiated ethanol at 150-190°K showed that the solvated electron was fully developed by the end of the pulse and decayed by a fast process followed by a slower process. These were attributed to fast decay of solvated electrons by recombination followed by a slower decay of free solvated electrons. From these observations, the yields of free solvated electrons and recombining solvated electrons were determined. A similar study in 1-propanol showed the same features above 180°K, and the yields of free solvated electrons and recombining solvated electrons were determined. The yields in 1-propanol were slightly lower than in ethanol.

Lower temperature studies showed evidence of a pre-solvated electron that decayed at the same rate as the solvated electron grew. Total solvated electron yields were measured from 150° to 220°K and were found to decrease with decreasing temperature. The rate constants for the slow decay of free solvated electrons were measured and they fitted an Arrhenius plot, with an activation energy of 20 kJ/mol.

Other Studies of Irradiated Liquids

The importance of the optical absorption maximum of the solvated electron in measurements of its yield and reactions led to an investigation of the nature of the solvated electron in various solvents. It was found that the maximum of the optical absorption spectrum of the solvated electron in liquids, including those containing the functional OH group (water and a number of alcohols), correlated well with the classical semi-continuum charge-solvent interaction energy over wide ranges of temperature and pressure. The results suggested that solvated electron cavity radius was predominantly determined by the polar group of the solvent and that the effect of alkyl substitution was secondary.

Magnetic Field Effects on Ion Recombination in Liquids

By irradiating samples in the centre of a magnet coupled to the 5-ns pulsed electron beam and fast optical detection system, Whiteshell staff was able to do ground-breaking research on the effect of magnetic field on ion recombination in irradiated liquids.

The initial products of irradiated organic liquids are positive ions and electrons. In the absence of impurities or additives, these recombine in pairs to produce excited states. Since the recombining electrons were initially in the singlet ground state of the solvent molecule, very fast recombination would result in only singlet excited states. Alternatively, very slow recombination would lead to a singlet-triplet ratio of between 1 to 1 and 1 to 2 in aromatic hydrocarbons in non-polar liquids. This discovery suggested that spin correlation is partly but not completely lost.

Initial experiments using fluorene in squalane as solvent were also done. Squalane is a viscous non-polar oil expected to slow down ion recombination sufficiently to test the validity of the spin correlation theory. The experiment was a success; the emission from singlet excited fluorene molecules was shown to increase in a magnetic field. Since the emission persisted longer than the natural lifetime of the fluorene singlet excited state, it was strong evidence that it arises from duplicate ion recombination, as predicted. Comparisons using solvents of different viscosities (squalane, cyclohexane, benzene) and different solutes (fluorene, anthracene, dimethylanthracene) confirmed the theoretical predictions. Similar effects were found in irradiated solutions of different solutes (fluorene, naphthalene, perfluoronaphthalene, biphenyl, perfluorobiphenyl) and solvents (squalane, cyclohexane, benzene, isooctane), adding further support to the hypothesis.

High-Temperature Solution, Solid Gas-Phase and Interfacial Chemistry

High-temperature water is an important constituent of CANDU reactors and heavy-water plants. Not only are high pressures required to keep water liquid at high temperatures, but also hot water is very corrosive.

Chemists at Whiteshell began to contribute to knowledge in this field in the early 1970s. The program was based primarily on chemical thermodynamics. This branch of the science deals with heats of chemical reactions, heats of solution, and other forces driving changes of chemical and physical state.

George Schultz - Applying chemistry concepts to CANDU design

The initial focus was on the thermodynamic variables of compounds of the metals present in process streams of reactors and heavy water plants (iron, cobalt, and nickel in particular). Several methods for extrapolating room-temperature data to temperatures up to 300°C were advanced. Electrochemical, solubility, and (later) calorimetric measurements were used to determine some of the

41

basic data to augment computational work.

As the waste management program advanced, the same techniques were used to predict the high-temperature properties of about one-hundred different chemical species of the actinides (primarily uranium, neptunium, and plutonium) in solids and high-temperature water solutions. The solubility of uranium dioxide was measured up to 300°C and the behaviour of glass waste forms in high temperature water was studied. In collaboration with other branches, the solubility of uranium dioxide (CANDU fuel) was evaluated under a wide range of temperatures and geochemical conditions, and the behaviour of glass waste forms was studied.

The techniques mentioned above were also used to determine the properties of compounds of some fission products, especially technetium and iodine, at elevated temperatures. The work on technetium compounds was of special scientific interest, because this element does not occur naturally in isolable quantities, but is produced by fission of uranium (and other actinides) in nuclear fuel. The fission products were studied for their relevance to waste management and reactor safety analysis. The research results contributed to the basic knowledge of a wide selection of chemical elements in high-temperature water, and were quickly applied in several AECL programs, They were also applicable to other Canadian industries, such as mining and pulp and paper.

Reactor safety interests also prompted a substantial effort on the properties of gaseous species of iodine and several so-called semi-volatile fission products (e.g., cesium, molybdenum, ruthenium, and tellurium). As with the solid and aqueous studies, this research encompassed both experimental studies and computational methods.

Research on solid oxides and oxide/water interfaces (colloid chemistry, electrochemistry, and associated surface analysis) followed a similar evolution, linked to important applied programs of the day. These included detailed studies of the dissolution mechanism of uranium dioxide, the oxidation kinetics of uranium dioxide in air and other gas mixtures relevant to dry storage of irradiated fuel, and the investigation of solid materials with potential to immobilize specific radioactive elements. Colloid chemistry research was tied initially to the transport of radioactive particles in reactor coolant circuits and, later, to the potential transport of insoluble actinides in a fuel waste repository. Research on the properties of strong acids and non-aqueous solvents,

Andy Vikis - Chemistry leads to answers

including concentrated solutions of thorium and uranium, linked to fuel reprocessing, was active for several years in the late 1970s and early 1980s.

Heavy Water Chemistry

Basic studies on two of the alternative methods to produce heavy water, the amine-hydrogen and the hydrogen-water processes, were undertaken at Whiteshell beginning in the late 1960s. Water soluble catalysts for the hydrogen-water process were investigated from a fundamental point-of-view as alternatives to the surface catalysts developed at Chalk River.

For a brief but exciting period in the mid-1970s, Research Chemistry and several other branches company-wide responded to a call for help in rectifying operational problems with existing Girdler Sulfide heavy-water production plants in Ontario and Nova Scotia. The more acute problem of foaming within the tall process towers was addressed by small-scale experiments that demonstrated the previously unappreciated, slight innate "foaminess" of hydrogen sulfide (the key process chemical) in water. The second problem consisted of corrosion of carbon steel surfaces, coupled to deposition of iron sulfides at various locations in the plant circuits. Published results of the ensuing research at Whiteshell continue to be cited by present-day researchers, especially in the oil and gas industry where sulfide corrosion continues to be problematic in the production and transport of petroleum products.

Analytical Science

As programs progressed and new ones arose, analytical methods advanced in parallel to meet new requirements at Whiteshell. With the growth of the fuel development program, faster methods evolved to help determine the energy output from test specimens and, in mixed fuels, apportion shares to the appropriate isotopes of uranium, plutonium and thorium.

Water and organic coolant in the WR-1 circuits were analyzed for the trace elements responsible for the generation of radioisotopes that were then carried outside the reactor shielding by the coolant. These methods were also used to analyze water in waste-management studies. Special attention was given to determining several elements simultaneously in the same sample or continuously in liquid or vapour streams.

Advanced techniques were used to analyze the surfaces and near-surface layers encountered in research on catalysts and corrosion. The chemicals used to decontaminate reactor circuits required analyses. The amine-hydrogen process for heavy water raised unusual requirements for the analysis of organic-nitrogen compounds. Investigation of pressure-tube hydriding required the determination of low levels of deuterium in many highly radioactive specimens cut from tubes removed from the Pickering reactors.

The mid 1960s was a time when a wide range of new analytical techniques were first commercially available, including GC-mass spectrometry, liquid chromatography, fast Ge-Li gamma detectors, in-line analyzers and scanning electron microscopy. As well, a large amount of research on the characterization of nuclear materials was being made available for the first time from the Atomic Energy Commission (AEC) of the U.S. and the United Kingdom Atomic Energy Authority (UKAEA). All these factors combined to provide great opportunities for developing the Analytical Science Branch (ASB).

In the 1970s, the rapid growth of the Nuclear Fuel Waste Management research program at Whiteshell required a significant increase in analytical resources, both human and mechanical. Automated

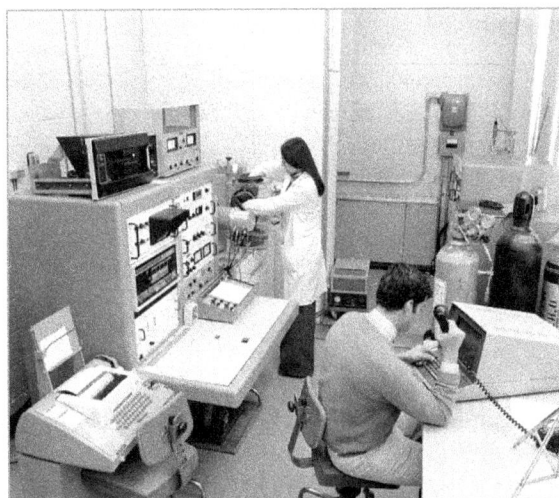

Stewart McIntyre and Annette Skeet

instrumentation was a key to advancing the research by generating large numbers of analysis results in a short time period. New equipment invariably included sample changers to allow unattended operation.

The "Super-Rabbit" automated NAA facility was the most impressive of these: a custom design with capabilities unmatched anywhere else in the world. After a brief but spectacular life analyzing rocks, biological material, and laboratory samples, it was laid to rest with the shutdown of the WR-1 reactor in 1985. Other computer-controlled instruments worked tirelessly in the Counting Lab and the Hot Cells to satisfy waste management researchers, developers of the dry storage concept for CANDU fuel, and environmental scientists. In parallel, multiple projects that generated commercial revenue included sensitive measurements aimed at detecting Chernobyl radioactivity in French wine and Canadian grains and oilseeds, as well as the Hazardous Materials Analysis Centre, serving the pulp and paper industry. These projects required pioneering the development of an accredited quality assurance program. Meanwhile, the fertile minds of the instrument development staffs produced a variety of sensors, for engine wear debris (Ferroscan), steam quality (Tundra), uranium ore grading (U-1000), and air quality (Comfocheck). Commercial revenue also allowed ASB to pursue experiments in "cold fusion" which led to several publications, including a note in *Nature*.

The biggest instrument success came in the 1990s with the Cerenkov Viewing Device development program. The hand-held instrument revolutionized the safeguards activities of inspectors from the IAEA at power reactors around the world. It brought ASB international recognition for the quality of the equipment and of the training program for IAEA inspectors, developed in collaboration with Canadian, Swedish and Finnish regulatory agencies. Instrument development, sales, service, and training also brought in revenues and led to the creation of a spin-off company.

George Grant and Dan Archambault

The ASB represented for some time the largest and most comprehensive analytical chemistry effort in Canada. As such, the Branch was asked to participate in major programs from sponsors such as the Canadian Armed Forces, AECL Power Projects, the International Atomic Energy Agency (IAEA), Ontario Hydro, Sask Power and Energy Mines and Resources in addition to handling thousands of individual analytical requests raised annually from within the Whiteshell site. Whiteshell was an agile participant in the race to support and improve CANDU reactor technology for over 40 years. However, the contributions of the ASB were not purely analytical or limited to laboratory analyses to support other departments. Staff were encouraged to conduct their own research projects in support of or tied to other site programs. Often they required unique experimental design, special facilities, and new analytical methods.

Analytical Science Branch – 1985: Back Row: Mike Ross, Al Hildebrandt, Paul Barnsdale, Andy Orr, Fred Doern, Bill Boivin, Barb Sanipelli, Darlene Hood, Andy Gerwing, Clarence Musick,Willy Dueck, John Montin, unknown, unknown, Eddy Lau, George Schultz, Darrell Hartrick; 2nd Row: Roy Taylor, Mike Attas, Terry Howe, Grant Delaney, Ahmad Solomah, Dennis Chen, Linda Brown, Dave Watson, Jim Betteridge, Dean Randell, Bonnie Bailey, Ken Wazney, Ela Rochon, Marv Arneson, Andy Kerr, Deb Brown, Nancy Pshyshlak; 1st Row: Rich Hamon, Unknown, Ernie Bialas, Karen Ross, Mike Lau, Bruce Lange, Randy Herman, Bill Kupferschmidt, Chuck Murphy, Keith Chambers, Bruce Stewart, Pat Ramsay, May Heinrich, Monique Chenier

ASB was constantly required to renew and revise its skills, equipment and knowledge base. In the early 1970s the activity transport program required increased capabilities for analysis of solution and solid products of corrosion in CANDU steam generators. Such skills were transferred quickly to support of the race to combat corrosion in heavy water plants. The nuclear fuel waste management program required characterization of many potential waste forms such as glasses, ceramics and fuel, as well as all the components of the repository. In the 1980s, worldwide nuclear safeguards programs gave rise to the development of remote detection methods for stored fuel. In more recent times, the major focus has turned to a monitoring of all aspects of the decommissioning of the WR-1 reactor. Armed with such varied experiences and skills, members of the ASB have made outstanding contributions to Canadian science and technology within many fields.

Fuel Reprocessing and Fuel Development Programs

In the 1960s and 1970s, consideration was being given within AECL to fuel cycles other than natural uranium, including fuels enriched with plutonium. The residual U-235 in natural U fuels after burnup was very low, so a process was examined to only recover Pu. The process involved dissolution of fuel in nitric acid, followed by the selective solvent extraction of the Pu with tertiary amines. The tests were conducted by the Chemical Technology Branch (CTB) using multi-stage mixer-settler equipment, and using a commercial amine known as Alamine 336. The analytical support included stream analysis for Pu and U in both aqueous and organic media using various titrimetric, colorimetric, x-ray fluorescence and radiochemical methods. Samples were prepared for alpha counting, alpha spectrometry and gamma counting. Absorption spectrometry was used to identify Pu oxidation states, gas chromatography and thin layer chromatography for amine and amine impurity analyses, and ion exchange and solvent extraction to treat samples for ongoing analyses. Much of this work was done in a glove box or a fume hood. Later work was performed with irradiated fuel and fission products using gamma spectrometry.

In the 1970s, thorium-uranium mixed oxide fuels were being considered for CANDU applications. ASB support was provided by conducting fuel dissolution rate tests in HNO_3/HF media. Samples of both irradiated and unirradiated fuel were subjected to a variety of acid concentrations and temperatures.

The effects of sample crushing and the Zircaloy sheathing were investigated. The thorium concentrations were generally determined by ethylene diamine tetraacetic acid (EDTA) titrimetric methods. Various complexation methods were used to determine the other components. For irradiated samples, the gamma-spectrometric analysis of Ce-144, a fission product, was shown to provide a good means of monitoring the dissolution rate.

When Whiteshell started to consider uranium carbide (UC) as a potential reactor fuel, ASB began studies of the hydrolysis reactions of UC at temperatures from 25° to 2500°C. The work was conducted with both unirradiated and irradiated fuel samples; hydrolysis kinetics and the overall reaction mechanisms were determined by measuring the gases produced. Sample surfaces were characterized using electron micrograph techniques prior to hydrolysis. Carbon contents were determined using combustion with oxygen. Oxygen content was determined using an inert-gas fusion method followed by gas chromatography. In general, the hydrolysis reaction products are UO_2 with varying ratios of CH_4, H_2, CO and CO_2, depending on reaction temperatures. A most interesting part of this work was to find a way to study reactions of molten UC at 2500°C, as there was no sample holder known that could withstand such temperatures, or that would not react with steam or the sample itself. A novel method of levitation of the UC samples using 45 kW radiofrequency induction coils surrounding the sample chamber was developed. The sample would lift off and almost instantaneously melt, at which time the steam valve was opened. The initial vigorous reaction was quickly self-limiting due to formation of an outer layer of oxide.

The work on irradiated UC samples also included a study of the radioiodine I-131 releases and a determination of the iodine chemical species. This work was carried out in the main hot cell facilities. The analyses involved the collection of gases on different filter media to selectively trap the different species, followed by gamma-spectrometric analysis. In some cases solvent extraction and precipitation methods were used to separate species prior to radiochemical analysis.

Neutron Activation Analysis

Stable isotopes capture a neutron and form isotopes. If these new isotopes are unstable, they will decay by emitting electrons or positrons and often one or more characteristic gamma rays. Typically, a sample is placed in a neutron flux for a specified period and then analyzed by gamma spectrometry. If there are interfering gamma rays, the sample can be dissolved and the isotope of interest isolated by chemical separation.

In the WR-1 reactor, each individual sample was transferred pneumatically through a conduit between the WR-1 reactor in Building 100 and the Analytical Science Branch (ASB) laboratories in Building 300 over a distance of about 130 m. The samples were encapsulated in low-density polyethylene capsules and placed in a cylindrical capsule commonly called a "rabbit". This system was installed in 1968. Magnetic sensors were used to indicate when a sample had been inserted into the irradiation site. The rabbits were made from ultrapure iron. Iron also captures thermal neutrons and the radioactive rabbits were collected in shielded containers after use. The capsules were reused after the activated iron had decayed substantially. The thermal neutron flux was 8×10^{13} n/cm^2/sec and the transfer time from the irradiation site to the ASB laboratories was an estimated 20 seconds. Heat from the WR-1 reactor was also a concern as the polyethylene sample containers in the capsules were unstable above 110°C. Temperatures were measured using a series of waxes with well-defined melting points.

The technique was ideally suited to determine trace concentrations in silicon carbide, a material that is resistant to most attempts to dissolve it, iodine concentrations at the parts per billion level in organic WR-1 coolant, and gold at the parts per million level in filters used to determine the SO_2/SO_3 ratios in plumes from fossil-fired power stations.

Material Characterization

ASB activities were initially dedicated to supporting research in the Chemical Technology and Fuel Development Branches along with analytical support of the operation of the WR-1 reactor. Later attention was devoted to understanding the micro-structural composition of corrosion deposits and related alloy degradation in CANDU heavy water reactors.

Most of this early work was done using an energy-dispersive x-ray detector on a scanning electron microscope (SEM). Corroded metals in the reactor core were deposited on surfaces in the nuclear reactor. Understanding such processes required analytical techniques that were very sensitive to the chemistry and composition of each surface involved. A program in the ASB was started to develop x-ray photoelectron spectroscopy (XPS) for studies of the activity transport process.

In the 1970s, as the nuclear waste management program became a dominant research activity at Whiteshell, additional surface analysis capabilities, Secondary Ion Mass Spectrometry and Auger Spectroscopy were added to improve the elemental sensitivities and measuring trace element distributions

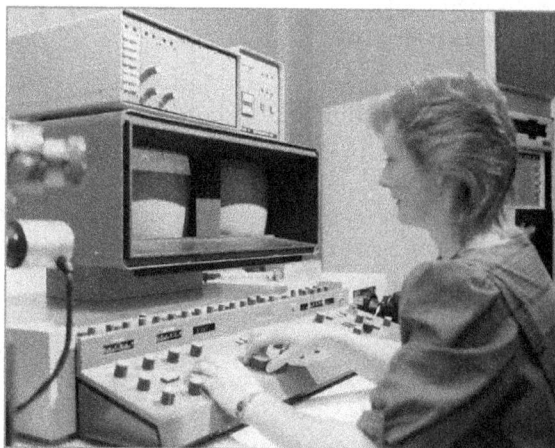

Karen Ross at scanning electron microscope

within prospective waste forms. In the 1980s came an increased interest in the microscopic characterization of fission product determination within irradiated fuel. This led to the installation of a "hot" SEM on the roof of the hot cells, allowing the measurement of cesium migration and other fission products within samples of the spent fuel. This work was followed by adaptation of the original XPS spectrometer to accept small amounts of spent fuel for characterization of the chemical states of fission products. This was one of the very first such facilities in the world.

Operation Morning Light

In the early hours of January 24, 1978, a Russian nuclear-powered satellite, Kosmos 954, made a fiery re-entry into the earth's atmosphere dispersing debris over the Northwest Territories. The Canadian

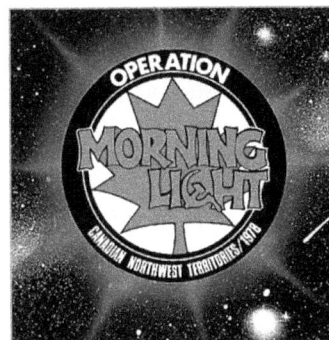

Government, with U.S. co-operation, promptly began a search and recovery effort known as Operation Morning Light. The ground search area extended 800 km from Great Slave Lake northeastward toward Baker Lake. Flights of high-attitude balloons in the polar stratosphere detected enriched uranium-bearing aerosols at concentrations and particle sizes compatible with a partial burn up of the satellite.

Canada's Department of National Defence was in charge of overall operations. Agencies affiliated with the Department of Energy Mines and Resources provided scientific and technical support. The Geological Survey of Canada conducted airborne radiometric surveys using techniques that distinguished between natural sources of radiation in the rocks in the search area and artificial sources from the reactor. The Atomic Energy Control Board of Canada was responsible for the safe recovery, transportation and storage of the debris. A team of scientists at Whiteshell received and characterized the debris to help assess their radiological and chemical impacts on people and the environment and unravel the type of reactor on board.

The spacecraft weighed approximately 4,000 kg and its nuclear reactor was thought to produce 100 kW or less of power. It contained about 50 kg of highly enriched uranium and a wide spectrum of radionuclides generated by fission of uranium atoms in the core and by neutron activation in the surrounding structural components. It also contained a number of chemically toxic elements. Many of the reactor components disintegrated during re-entry and underwent chemical transformations. The debris was distributed over an area of thousands of square kilometers and the threat to human populations, wildlife and the quality of water and soil became a concern. Over 10 months, crews recovered more than 4,000 pieces of radioactive debris, about 65 kilograms in all, from the Northwest Territories and northern Alberta and Saskatchewan.

The first shipment of space debris arrived at Whiteshell on February 5, 1978. Recovered fragments and hundreds of tiny particles of radioactive dust were examined in hot cells and subjected to physical, chemical, metallographic, microstructural, radiochemical and isotopic analyses. Many of the particles were detectable only by radiometric surveys and retrieved by shovelling snow into bags. In all, 15 major shipments were received; more than 40 scientists and technicians were involved; and 4,700 analyses were reported.

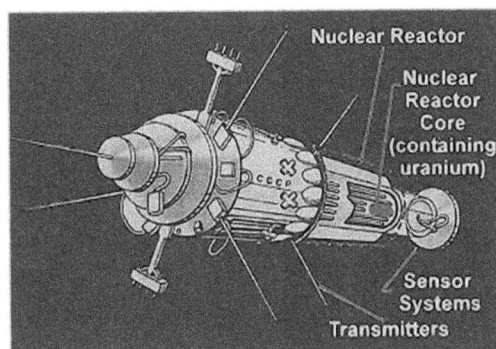

Kosmos 954

The recovered debris generally fell into the following general classifications:

- Fragments of steel plates often covered with slag and contaminated with mixed fission and activation products.

- Beryllium rods, contaminated with mixed fission and activation products, some with niobium cladding and scorched surfaces.

- Solid beryllium cylinders, some contaminated with

mixed fission and activation products.

- A complex assembly of six cross-connected tubes or legs and a circular base plate, nicknamed "the antlers". The exterior surfaces were covered with hydrated lithium hydroxide. The cross braces contained both lithium hydride and lithium hydroxide. The legs contained a black powder identified as a borosilicocarbide.

- A large nonradioactive cylinder resembling a "stovepipe" and composed of mild steel.

- Assorted radioactive chunks, rods and slivers.

- Radioactive particulates, usually sub-millimeter in size, with widely variable density, structure and composition. The high-density particles comprised uranium with about 10% molybdenum; the low-density ones consisted of a uranium-beryllium intermetallic compound.

These investigations helped determine the physical and chemical properties, structure and composition, radiological characteristics, and behaviour of the debris that survived re-entry. The data helped reactor physicists identify the reactor type, fuel composition and power history; aided materials scientist to identify alloys used and, in some instances, their method of fabrication; assisted health physicists to assess the chemical and radiological hazards through external exposure, ingestion or inhalation; and made it easier for field personnel to mitigate the risks to human health by removing radioactive debris from populated areas.

The main conclusions drawn from these investigations were:

- The reactor core disintegrated on re-entry, the nuclear fuel melted and reacted with surrounding components, and the fall-out zone was showered with fragments and of debris and a myriad of microscopic radioactive particles of diverse composition.

- The nuclear fuel was a metallic uranium-molybdenum alloy, highly enriched in ^{235}U, typically used in small low powered reactors.

- The intact beryllium rods and cylinders were probably used for reactivity control. The cylinders survived re-entry in pristine condition because they were protected by a refractory surface coating of beryllium oxide. Beryllium is a neutron reflector. It is chemically toxic.

- The activation products identified in structural components were produced by high-energy neutrons that passed through the beryllium reflector.

- The copious amounts of lithium hydride associated with the "antlers" were probably associated with the reactor control system. The boron found within the tubes is a strong neutron absorber that was used for reactivity control.

- The activation products and fission products found in the steel components helped determine the types of steel used in construction of the satellite.

The crash of Kosmos 954 raised international policy questions. Soon after the satellite's crash, there was a call from the United States to prohibit satellites containing radioactive material from orbiting the earth. This was followed by similar calls from Canada and countries in Europe. In November 1978, the United Nations authorised its Committee on the Peaceful Uses of Outer Space to set up a working group to study nuclear-powered satellites.

In the months following the crash, the Canadian government sought compensation from the Soviet Union under the 1972 Convention on International Liability for Damage caused by Space Objects. This international law holds satellite owners liable for the damages caused when space objects fall back to earth. The Soviets fought this case claiming that Kosmos 954 had broken up by the time it fell to earth and thus could no longer be recognized as a "satellite" when it landed in the Northwest Territories. The fall of Kosmos 954 not only established an occasion to test satellite liability law, but Operation Morning Light became a prototype for future satellite recovery missions.

Medical Biophysics and Life Sciences

Research into the life sciences was directed towards the biological effects of radiation, the behaviour of radionuclides in the environment, the use of radionuclides for biological research, and developing instruments for radiation protection. Mathematical models were developed to describe the metabolic behaviour of radionuclides and the resulting doses from them to organs and tissues in the human body. The effects of radiation were studied not only on complete organisms but also at the cellular and molecular levels. Important information was obtained on the association between variations in radiosensitivity, DNA repair deficiency and cancer proneness in humans; an understanding of the various levels of radiosensitivity in people with cancers is necessary to establish a sound protocol for their treatment by radiation.

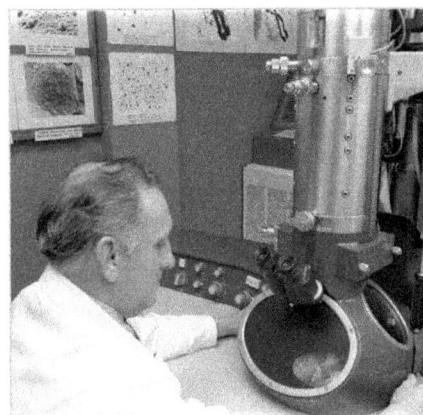

Ted Copps at scanning electron microscope

Medical Biophysics

Research in the Medical Biophysics Branch was divided into the four categories: Radiobiology, Radiation Biochemistry, Biochemical Technology, and Molecular Radiobiology.

The Medical Biophysics group at Whiteshell worked on the molecular biology of DNA by researching early events occurring in fractions of a second following the deposition of energy into the system. These early physical events included, for example, the production of free radicals, which cause chemical damage in DNA and in other biological constituents of cells. The chemical mechanism of the interaction between radiation and chemical carcinogens was investigated. The role of naturally occurring enzymes, such as superoxide dismutase and peroxidase, was studied to learn how these enzymes protect biological structures from being damaged by the highly reactive free radicals. It was found that these enzymes degraded the potent free radicals to less toxic molecules. This work has greatly improved our understanding of how radiation produces biological changes in living organisms.

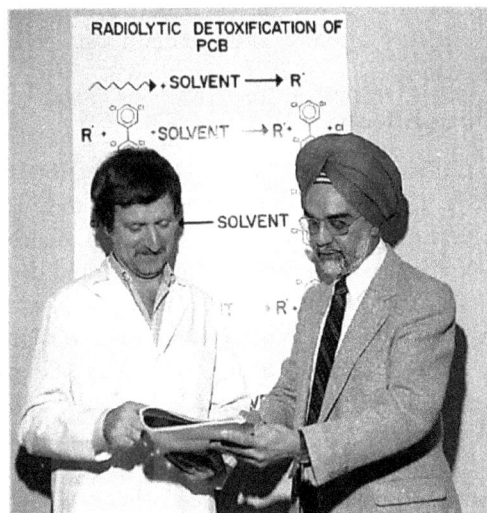

Walter Kremers and Ajit Singh

The mechanism of action of chemicals, known as radiosensitizers, which increase the sensitivity of cells to the lethal effects of radiation, was also investigated. The ultimate aim of these studies was to improve the radiotherapy of cancers. The lack of cures for and the recurrences of cancers were believed to be due to the fact that cells in the centre of tumours were not killed because they are anoxic or lack oxygen, and for this reason were radiation resistant. Although radiosensitizers were effective in counteracting the radiation resistance of anoxic cells, cancer treatment centres found that radiosensitizers had considerable toxic side effects. The sensitivity of mammalian cultured cells to heat and radiation was also studied. Results of these studies had implications for the use of hyperthermia in the treatment of cancers. Preliminary promising therapeutic results were obtained on clinically terminal cases.

Another focus of research at Whiteshell was work on membranes. Membranes are important in cell functions, because they control the exit and entry of important food stuffs, waste products, minerals and water, while retaining macromolecules such as DNA and proteins. For this reason it was suggested that damaged membranes could lead to delayed effects (cancers and genetic defects).

One of the scientists at Whiteshell used model membranes prepared from biological materials to study the integrity of the membranes after exposure to low doses and low-dose rates of radiation, as well as to compare the effects of x-rays with those of tritiated water. From his studies, he claimed that as the dose rate was decreased, the effects on the model membrane per unit dose increased.

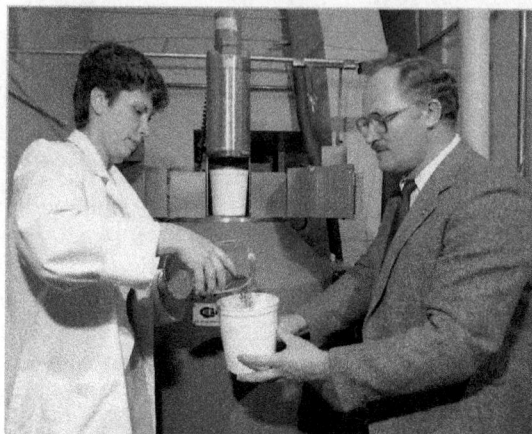

Noemi Chuaqui-Offermanns and Jos Borsa

This observation, which received some publicity in the popular press, is at times called the "inverse dose rate effect," where the effectiveness of low-dose rates is greater than higher dose rates. The interpretation of these data generated considerable controversy within and outside AECL. Most of the AECL scientists did not accept the results or the interpretation of the experiments. In 1980 the United States National Research Council's BEIR committee reviewed this work and concluded that "although it is well recognized that membrane integrity is essential for normal cell function, there is inadequate basic understanding of membrane structure and function on which to base a detailed theory of radiation-induced damage mechanisms."

Petkau Effect

In 1972, a Whiteshell researcher, Dr. Abram Petkau, found that when a cell membrane was irradiated slowly, a smaller total dose was needed to cause damage. Petkau had been measuring the radiation dose that would rupture a simulated artificial cell membrane. He found that 3.5 krads delivered in 135 minutes would do it. Then Dr. Petkau repeated the experiment with much weaker radiation and found that 0.7 rad delivered in 690 minutes also ruptured the membrane. This was counter to the prevailing assumption of a linear relationship between total dose or dose rate and the consequences. The ionizing radiation produced negative oxygen ions (free radicals). Those ions were more damaging to the simulated membrane in lower concentrations than higher because in the latter, they more readily recombined with each other instead of interfering with the membrane. The ion concentration directly correlated with the radiation dose rate. This effect came to be known as the "Petkau Effect".

Abe Petkau

Petkau also found in 1976 that the enzyme superoxide dismutase protected cells from free radicals generated by ionizing radiation, obviating the effects seen in his earlier experiment. Petkau also discovered that superoxide dismutase was elevated in the leukocytes (white blood cells) in a sub-population of nuclear workers occupationally exposed to elevated radiation; further supporting the hypothesis that superoxide dismutase is a radioprotective agent. Thus, Petkau's original 1972 experiment revealed the potential effects of ionizing radiation on cells without natural radioprotective mechanisms in place.

Environmental Research

Environmental research at Whiteshell focussed on the transfer of radionuclides from soil and water to plants and animals. Notable experiments included spiking a swamp with iodine, measuring radionuclide migration in soil columns (lysimeters) and using ping pong balls to measure gas discharge from buried rock fissures. Sampling was done across Canada, including the arctic. In addition, the scientists studied non-nuclear topics, such as PCB behaviour and odd exposure pathways such as people inadvertently eating soil. In 1989, Whiteshell staff played a key role in writing the IAEA Handbook of Radionuclide Transfers, including providing all the soil data, some of which is still the standard for use globally.

Dennis Thibault and Marsha Sheppard - environmental sampling

As part of the overall environmental program, it was important to obtain experimental data on the dispersion of radioactive material released into bodies of water. Since some of these bodies of water were public areas, fluorescent dyes rather than radioactive materials were used as tracers. Such dispersion studies were made in the Ottawa River at Chalk River, in the Winnipeg River at Whiteshell, and in Lake Huron at Douglas Point. The data were used to establish dispersion models at these various sites. The concentrations of tritium measured downstream in several places in the Ottawa River after an accidental release of tritiated water from Chalk River in 1988 was in good agreement with the values predicted by the model.

Bill Evenden and Steve Sheppard – Uptake of radiation by local plants

Environmental Monitoring

AECL has always provided personnel dosimetry services and radiological monitoring at their sites, and report regularly to their regulators. The routine environmental monitoring program has been continually optimized to address parameters of potential concern, to confirm that appropriate mitigation measures are taken, and to assist in the development of appropriate responses to unforeseen events.

The major operating facilities were sources of radiological effluent releases from the site. The primary source of liquid effluent releases was the process water outfall, which discharged continuously to the Winnipeg River. The secondary source of liquid effluent was the sewage lagoon, which was normally discharged twice per year to the Winnipeg River.

Annual measurements of radiation levels and radioactive contamination within and outside the WL site boundary have been performed since the 1960s. These measurements verified that levels of radiation and radioactive contamination due to

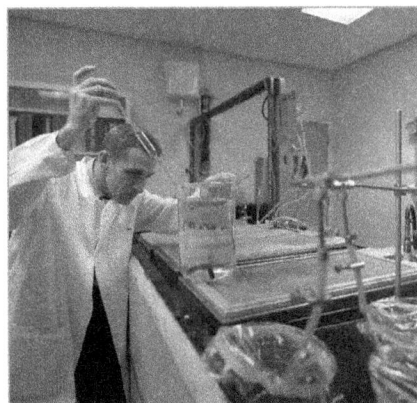

Orville Acres

operations at the site, as well as the resulting radiation doses to members of the public, continue to be below regulatory limits and guidelines.

Monitoring of potential liquid effluent exposure pathways have confirmed small contributions from WL operational and decommissioning activities. Cs-137 and Sr-90 appear in downstream samples of Winnipeg River water, vegetation and fish. Radioactive contaminants in Winnipeg River water have always been very small fractions of allowable levels.

Monitoring of potential atmospheric effluent exposure pathways has not indicated any measurable dose contributions from the site activities. This is also consistent with effluent monitoring results, which indicated that airborne emissions were very small.

The estimated dose to members of the public due to radioactivity in WL effluents, based on the annual environmental monitoring results, have always been very small compared with the regulatory public dose limit and with doses to the Canadian public from natural background radiation.

Dean Randell and Tavis Donnelly – Second Generation Researchers

A Fitness-For-Service (FFS) assessment was completed in the 2008 year for the WMA. This assessment confirmed WMA ability to provide adequate containment of the stored waste for a period of not less than 30 years. In addition it provided recommended maintenance measures and long-term monitoring requirements. Ongoing radiological surveying and monitoring was continued at the perimeter of the WMA fence to confirm that baseline conditions were not changing.

Monitoring of the hydrogeological conditions at the WMA was also done. Monthly water level measurements are taken for all groundwater-monitoring wells at the WMA, and hydrographs of water level versus time were constructed and analysed. The work provided confirmation that the primary hydrogeological conditions at the WMA remain as expected.

AECL continues to be committed to maintaining an environmental monitoring program for as long as wastes requiring management remain at the site. They also involve local communities in the

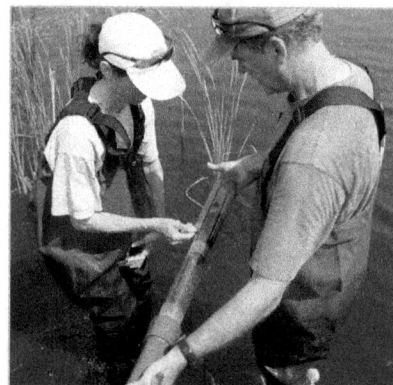

Martha Laverock and Dean Randell - environmental sampling

environmental monitoring program and implement mitigation measures for project activities as needed.

Whiteshell Organization and Services

This section touches on the organization and the service people that provided support to the scientists during the OCR and other research programs undertaken at WL. Considerable effort was made to recall past staff within these service groups, provide as many photos as practical and allude to their program involvement.

In the fall of 1963, WNRE and the Pinawa town site still had a large backlog of nearly finished construction. The 100-ton bridge over the Winnipeg River was completed and Highway 211 was gravelled. At the time, the bridge had the highest capacity of any bridge built in Manitoba. It would handle the massive flasks and trucks that would transport shielded irradiated fuel and radioactive wastes to and from the WNRE site. Even now, in 2016, the bridge has the capacity to handle any loads expected from the decommissioning activities at WL.

The WNRE site housed many of the contractor's staff in temporary bunkhouses. A large temporary cafeteria located near the bunkhouses provided three massive meals a day and off-hour lunches too. This cafeteria continued to operate until a permanent cafeteria was built a couple of years later but the meals were never as large as they were when the contractors were around. Crawley and McCracken and Dominion Catering were some of the contractors that provided food services to the site. Located near the bunkhouses was a sheet of outdoor curling ice with a set of rocks that provided entertainment for contractor staff during the winter on their off hours. These rocks were moved from the plant site to the town site in the early fall of 1963, where one sheet of ice was built and maintained that winter along the Southeast corner of the F.W. Gilbert School yard.

F. W. (Fred) Gilbert was in charge of managing the contracts associated with building and operating the plant and town sites. He selected staff experienced in the fields of project management, contract management and procurement, which had relevant experience with consultants and contractors. He also hired building inspectors, administrators, finance, public relations, security, firefighters and human resource personnel.

Pre-1963 Pioneers
Max Allan
Bud Bjornson
Cy Crawford
John Fenton
Fred Gilbert
Irv Grabke
Major Hammond
Gerry Hampton
Bill McEwan
Dave Morgan
Jim Putnam
Harry Smith
Roger Smith
Shawinigan Engineering
Wardrop Engineering

Prior to the fall of 1963, he and his staff worked from offices located in Winnipeg. However, as construction progressed and housing and services became available at the town and plant site, staff began a slow migration from Winnipeg to Pinawa. By the end of August 1963, the initial wave of families had established themselves in Pinawa or in adjacent communities. Most of these people were AECL employees, but some were contractors, consultants and public service employees. As an example, early residents of the town site included: one RCMP officer; a few Shawinigan Engineering employees who were managing site construction projects; post office employees; doctors, nurses and hospital administration staff. In many cases, these people were accompanied by their families.

Meanwhile, in Chalk River, the planning of the scientific program was well under way and many of the researchers who would lead the OCR program were planning their experimental and staffing requirements in preparation for their scheduled move to WNRE in 1963. Staff hired during this early period first moved to CRNL and immediately became involved in their part of the research program. Some experimental facilities were constructed and commissioned at CRNL and transported to WNRE for installation, thus allowing some researchers to begin work soon after their arrival to WNRE. Some of the WR-1 reactor operation personnel were hired and housed at CRNL for training prior to their move to WNRE. They became familiar with reactor operations, provided some oversight on the design and produced reactor-operating manuals. Experienced reactor operators were transferred from research reactors at CRNL to the WR-1 reactor group. Other WR-1 reactor operators were hired after the senior operators moved to WNRE. Most of the researchers moved from CRNL to WNRE during August, before the start of school in 1963. The reactor operations personnel moved later, in 1964, when construction of the WR-1 reactor was further advanced.

The initial two years, or so, were amazing. The site organization was completely fluid. The construction of buildings and infrastructure was nearing completion. Experimental facility construction and commissioning was accelerating. The staff was working from unfinished offices, boxes were used for desks, any chair would do, dust was everywhere, the toilet facilities were usable but not partitioned, few phones were available, but it was the best of times! Social boundaries between trades, scientists, engineers, technologist, technicians, draftspersons, secretaries and others essentially disappeared during this period.

At the end of this period, test facilities proliferated and researchers were busy but something subtle had occurred. Our fluid organization had congealed and had become formal. It now consisted of a "site head", several "divisions" with two or more "branches" and in some cases; branches had one or more "sections".

A typical organization chart (circa early 80s) is presented in the Appendix (see Chapter 7). In earlier years, the chart was less complex but always used a vertical flow of authority. In the mid-eighties, it became a matrix where authority could come from other sites, divisions or branches.

It is important to acknowledge those that supported the scientists and kept the research reactor, plant buildings and the overall plant infrastructure operational at WL. These groups included trades people (electricians, millwrights, pipe fitters, welders, instrument mechanics, etc.), reactor operators, nurses, engineers, accountants, technologists, technicians, radiation surveyors, labourers, secretaries, telephone operators, librarians, power engineers, guards, firefighters, draught-persons, mail and purchasing staff, stores people, drivers, public affairs and many others.

A review of a typical WNRE Organization Chart (see Chapter 7) shows that four columns of the eight at WL represented "Service Divisions" with their branches. One of the larger service divisions was "Engineering Services", currently called Engineering Design and Operations. It included Reactor Operations, Nuclear Technology, Design and Project Engineering and, Maintenance and Construction branches. In the early eighties, the Maintenance and Construction branch became a division.

Reactor Operations Branch

WNRE had a mandate to perform research and development on a new reactor type, one using organic coolant with a heavy-water moderator. A research reactor called WR-1, located in building 100, was built with a maximum thermal power design capability of 60 MWt, although, when initially built and commissioned it could only operate at 40 MWt because the outer circumference of the core was intentionally left without a full complement of fuel channels. These were added several years later.

It only took a hand full of qualified people to operate the reactor on a twenty-four hour basis, but a significant number of research people were required to plan and analyze the results of experiments placed in the core of the reactor. The outcome of these experiments, either positive or negative, affected future designs of commercial organic-cooled power reactors. Although the OCR had many features that were better than the heavy-water cooled and moderated reactors and many design studies were completed, no prototype organic-cooled, heavy-water moderated reactor was built in Canada.

Eric Graham and Dick Meeker

The reactor building was one of the most interesting places on the site. It not only contained the WR-1 reactor, it was also the focus of many research projects, especially those relating to reactor materials, organic coolant, fuel development, zirconium metals, optimisation of coolant composition and filtration systems, to name a few.

When one walked into the reactor building, one never knew what to expect. If the reactor was operating, the building would be reasonably quiet; a few operators, maintainers, instrument mechanics and shift supervisors would be around. Most of the activity would be in the control room or reactor hall where operators and shift engineers were monitoring instrument panels, clearing alarms and keeping up with paperwork. The faint smell of the organic coolant was always in the air.

If one walked in when the reactor was shut down, it was a beehive of activity. Scheduled maintenance was only possible when the reactor was shut down. Operators, radiation health surveyors and trades people would be working in locations with limited access or no access during reactor operation due to radioactive fields and contamination. During reactor shutdowns, site staff and off-site visitors were discouraged from entering the reactor building.

More than one visitor had their shoes, shirt, pants or other personal belongings contaminated with small amounts of low-level radioactive material. These items were usually returned after decontamination. However, this issue was easily avoided by the practise of changing your clothes and dressing into "whites" before entering the potentially contaminated areas of the reactor. The term "whites" comprised white overalls, socks, gloves, cotton head cover and shoes, supplied by AECL. If your body became contaminated, you showered until you were clean. Special precautions were required (e.g., respirators and special outerwear) if there was a chance of inhaling or ingesting airborne contamination.

A shift complement of between 9 to 11 people usually consisting of a shift engineer, trainee shift engineer, shift supervisor, shift lead hand, 3 or 4 operators, an instrument mechanic, and a radiation and health surveyor were required to keep the reactor operating 24 hours a day, seven days a week. Initially, these personnel were scheduled on three eight-hour shifts. In the early 80s this was changed to two, twelve-hour shifts per day, an arrangement that allowed the personnel to work four-days with three-days off each week. The twelve-hour shift turned out as the favourite by most.

George Scharer and Kevin Rogers

Shift engineers and trainees had an engineering degree, usually in a discipline like chemical, electrical, mechanical or engineering physics. The other operations shift personnel had a variety of backgrounds, typically from trades, technical schools or positions in process operations and maintenance. They were certified after successfully completing an extensive on-the-job training program.

Not all reactor personnel worked on a shift schedule. Administrative and technical staff such as reactor physicists and supervisors and their support staff worked 7.5 hour days, five days per week. Most were on call though, in the event they were required at the reactor during off-hours.

Operations Branch – 1st Row: Jack Middleton, Barry Hood, Barb Borgford, Ron Oberick, Dave Tymko, Frank Oravec, Lloyd Rattai, Lorna Reschke, Bun Baxter, Les Hembroff; 2nd Row: Don Bruneau, Jim Clarke, Vic Reschke, Glen Snider, Larry Gauthier, George Snider, Cliff Zarecki, Ray Webber, John Kerr, Vern Steiner, Gaylord Newman; 3rd Row: Bob Bruneau, Blake Cutting, Roy Ticknor, Tom Barnsdale, Vic Popple, Gerry Plunkett, Mike Berry, Wally Dobush, Harold Bender, Gary Simmons, Thor Borgford, Nat Fenton, George Scharer, Al Jarvis, Tom Tabe, Dave Clark, Carl Sabanski

The WR-1 reactor personnel were also responsible for the operation of the Active Liquid Waste building. The water coming from Building 100 that might be potentially contaminated with radioactivity or organic coolant (e.g., from active area sinks and showers and from other decontamination activities) was directed to the Active Liquid Waste Treatment building and placed in storage tanks for sampling and treatment prior to release to the WMA or the domestic sewer system.

At one time, during the height of the reactor experimental program, a plan to replace the shift engineers with engineering technologists took place. The replacement technologists were hired and were in training when the program was terminated. It turned out the AECB disapproved of this arrangement and the concept was dropped. However, most of the technologists in this program were quickly absorbed in the general R & D programs at WNRE, while others re-established themselves in non-nuclear fields.

Design and Project Engineering Branch

The mandate for this branch included managing major site projects, providing a complete design and drafting function, along with a non-destructive testing (NDT) inspection service. The branch also provided secure storage for all engineering and drawing records and developed and maintained site engineering and drawing standards.

The Engineering Design Section was first established and located in the basement of building 300. Walter Litvinsky was the first designer-draftsperson hired. Jim Gold was hired as Chief Engineer. This was the start of the Design Branch. Around 1967, a 3-storey administration building was built to provide office space. Space was assigned to the Design Branch and various other groups.

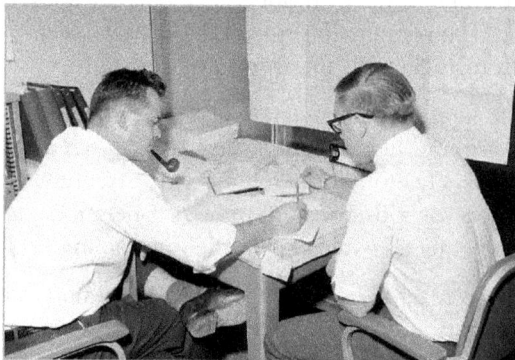

Ray Sochaski and Doug Benton

The Design Branch staff levels quickly stabilized at about 80 people. About 20% were engineers, 65% were design-drafting staff, 5% clerical staff, 5% inspectors and 5% were tracers. The design-drafting staff was subdivided in sections of mechanical, electrical, and civil disciplines. Clerical people provided secretarial services to the branch and support to the drafting and engineering staff with blue printing, record storage, filing and retrieval of prints. They also maintained records on each job passing through the office. Each job record contained design calculations, correspondence, blueprints and other relevant information.

In the early 60s, drawings were produced on drafting-tables that could be rotated between the horizontal and vertical and levitated up and down allowing work to be performed in a sitting or standing position. These tables were located in open spaces and when extended, they looked like an armada with sails. In the 1980s, they gradually disappeared; computers replaced them! In the beginning all engineers, designers, draftspersons and inspectors within the branch were men. Clerical people were a mix of men and women and tracers were mainly women. By the mid-seventies, women were starting to hold a few of the engineering and drafting positions.

Larry Ramsay, Wayne Melin, Dave Litke, Jack Reimer, Garry Schellenberg

In the pre-computer era, it is interesting to note that "blueprints" were characterized by a blue background with white lines and printing. These later changed to "whiteprints", with a white background black lines and printing. The variety of transparent paper available in the early days to prepare "master drawings" had several issues, such as being brittle, hard to correct, and difficult to ink and to store. This material was replaced with plastic Mylar, which greatly reduced these difficulties. Mechanical pencils replaced wooden pencils; motorized erasers replaced manual erasers and calculators replaced slide rules; simply destroying the slide rule industry in its wake. By the mid-80s, even the drafting tables and their drafting-machines were totally replaced by computer aided drafting (CAD) systems.

The Design Branch undertook the design of experimental equipment for the WR-1 reactor. Since WNRE was a licensed nuclear site and the license imposed certain conditions, the WR-1 Operations Branch reviewed and approved any changes to the WR-1 reactor operation, or its configuration. Since there were many experiments awaiting installation in the reactor during its operational life, a formal approvals process was established. The proponent would document the objective and requirements, including changes to the reactor's operation and configuration. A committee would review this documentation for impact to the overall site program and for budgetary acceptance. If approved by this committee, the documentation would go to the Design Branch for a cost estimate and formalization. During the design stage, continual discussion with reactor operations, reactor physicists, and maintenance

Walt Litvinsky

personnel occurred, which invariably initiated design changes. It also allowed Operations and the Reactor Physics section time to evaluate the effects the experiment may have on reactor operation and core reactivity, and it allowed maintenance to schedule their construction resources.

When final design approval occurred and the drawings were approved, construction would usually begin. Once an experimental facility was installed, commissioned and running, it was the practice to "as-built" all drawings to reflect any field variations to the approved drawings. The Design Branch was responsible for this final phase and tried to insure a minimum of "as-building" through close collaboration with operations and maintenance branches during the design, construction and installation phases of each project.

In general, the process worked well. Between November 1965 and May 1985, the period over which the WR-1 reactor operated, no serious events occurred to breach the limits of the site license.

Many other experimental facilities not related to WR-1 were also required at the site and the Design Branch was usually involved. Some, like the concrete canisters and experimental inserts for the hot cells, saw exposure to highly radioactive materials. Other impressive experimental facilities included loops that mimicked reactor conditions (e.g., RD-12, RD-14) to study loss of coolant events and the Combustion Test Facility that simulates hydrogen explosions, similar to that which occurred recently at the Fukushima Daiichi Nuclear plant in Japan. Other design initiatives included the addition of a small amount of heat to the plant water intake to prevent

Jim Gold

frazzle ice blockage during the Winnipeg River freeze-up and the URL construction to assess the potential for the permanent storage of used nuclear fuel in stable Canadian Shield bedrock.

WNRE also used outside engineering and architectural consultants when required. The Design Branch expedited these contracts through its project management group. Usually when a consultant was involved, a general contractor and many subcontractors would be involved too, and the cost of the completed project would be in the low millions of dollars. Experienced project managers kept these jobs on schedule and within budget. This was not an easy task; it required detailed drawings, good communication and contracts between consultants, contractors and WNRE. Once work started, it was the project manager's responsibility to hold "extras" to a minimum and maintain good relations between all contract parties.

The Design Branch was also responsible for providing trained inspectors to verify that in-house and contractor-built projects complied with drawing specifications, and international, national, provincial and internal AECL codes and standards. Typically, all reactor-related piping systems required their welds to be inspected using gamma and/or x-ray techniques. Components such as pressure vessels and heat exchangers required provincial certification. Piping, pump and valve manufacturers sold certified products to meet temperature, pressure and specified

Art Gauthier, Denis Godin, Marv Cooper, Greg Gowryluk, Bill Baker, Garry Stokes, Grant Miller, Erwin Schatzlein

operating conditions. Most inspectors employed by WNRE were fully qualified to interpret gamma and x-ray film, and to perform dye penetrant, eddy current, magnetic-particle and ultrasonic testing.

Engineering Branch in 2013: Front Row – Ken Scott, Christel Aitkenhead, John Coleman, Judy Clark; Back Row – Charene Davis, Ron Serediuk, Stuart Parrott, Eric Zhou, Alex Drivas, Josh Carolan, Muhammad Umair, Andrew Graham, Sana Rasheed, Syed Khan, Patricia Pawluk, Kevin Barager, Daisy Xu, Bruce Orbanski, Pam Meeker, Kamil Malek, Amina Radyastuti, Fuping Liu, Mike Enns

In 1970, Canada saw the approval of metrification and the initial phases of its implementation by the Liberal government under Pierre Trudeau. The system adopted was the "Système international d'unités" or International System of Units. The Design Branch was given the responsibility to incorporate it into the AECL/WNRE standards for immediate use. These times turned out to be most interesting.

Our scientists and researchers were already using a system of units very close to the recommended SI system adopted by the government and saw little difficulty in complying. On the other hand, the trades and many others never used metric measurements. The transition from the English to SI system of units for the trades and others was difficult and acrimonious.

Maintenance and Construction Branch/Division

This branch is one of the largest branches on-site. Several buildings were required to house all of their equipment and activities, but most of their operations were conducted in or from Building 412, located in the active area, and Building 408, located in the non-active area. Originally a branch of the Engineering Services Division, in about 1980, it was changed to a division called the Maintenance and Construction Division. It was home to most of the trades working at WNRE. These trades included welders, machinists, electricians, millwrights, instrument mechanics, tool and die makers, pipe fitters, carpenters, drivers, labourers, automotive mechanics, sheet metal workers, air conditioning mechanics, heat and frost insulators and other trades.

Housekeeping Maintenance

A considerable amount of effort was required just to maintain the site through the four seasons of the year. Lawns were mowed, trees trimmed, flowerbeds weeded and site property improvements were conducted mostly during the spring and summer periods. During the fall and winter periods, plants were winterized and protected from animals. Every weekday required the cleaning of buildings; washing, waxing and polishing floors, removing waste, and periodically cleaning windows. Many people were required to undertake this work. In some cases, local contractors were asked to bid on performing this work and over the years established reputable businesses providing these services.

This group also managed and operated site facilities for the placement of non-radioactive solids and liquid domestic wastes. These include office waste, maintenance and construction wastes and experimental waste. The solid and liquid wastes were placed in on-site landfill and sewage lagoons prior to release to the environment. This group, in association with radiation and industrial safety personnel, were also responsible for packaging and storing solid and liquid wastes with radioactive contamination. Wastes of this type were stored at a designated low or medium level radioactive waste site in various containment structures.

The Maintenance Branch also maintained a fleet of vehicles and drivers. As an example, a vehicle would leave the plant site twice a day to deliver mail generated at the plant site to the Pinawa post office and return with mail for the plant site. They might also perform secondary errands like deliver files and mail to or from the hospital, Kelsey House or other WNRE offices in the Pinawa town site. A car from the plant was also always available in the town site in case of a plant emergency. It was there to transport emergency staff to the plant site. There were also larger vehicles that performed heavy work on and around the site. Some moved dirt, gravel, snow and garbage, while others moved heavy flasks containing radioactive materials around the site, between WNRE and CRNL, and in some cases to and from the USA.

Cars were available to pick up visitors at the Winnipeg Airport. Staff was allowed to requisition cars and on many occasions; visitors were transported between the plant and the URL, the town site or other places in Manitoba.

Right from the very beginning of site operations, travel between WNRE and the Winnipeg airport was heavy and remained so for many years. During this period, various contracted local taxi services provided transportation for travelling WNRE staff. Some that come quickly to mind were the Augustine Brothers,

Joe and Emil, who owned a garage and Chrysler dealership in Lac du Bonnet; Tom Melnick, GMC dealership in Whitemouth; Lac du Bonnet Taxi, owned and operated by Romeo and Eleanor Hladki and possibly one or two others. Tom Melnick also owned and operated a bus service that shuttled staff between Pinawa and the plant site each working week day morning and evening.

Mechanical and Construction Branch - Front Row: Larry Novakowski, Harold Malkoske, Jack Turner, Ray Wazney, Mary Ryz, Len Molinski; Back Row: Gerry Lange, Don Zetaruk, Len Williams, Luke Garneau, Liz Hannon, Cliff Zarecki, Torrie Visca, George Snider, Sandy Mathews, Kathy Woodbeck, Ad Zerbin, Mike Wayne, Paul Sansom, Erwin Hemminger, Elmer Smith, Oswald Meyer

To look after WNRE's fleet of vehicles, a 3-bay garage in Building 408 was available for maintenance. The fleet consisted of the same make of vehicle, with spare parts kept in the WNRE Stores. Initially, WNRE staff provided the maintenance to these vehicles. Vehicle maintenance was contracted out in the 1990s but was still performed in the Building 408 garage. During this period, even privately owned vehicles were also accepted for maintenance. Eventually this work was moved out to local garage in the Pinawa area.

The Trades

The WR-1 reactor operated at a high temperature and pressure and produced power by the fission of uranium. The site was licensed based on providing safety to the local population and full disclosure of any changes in radiation level at site and the surrounding area. Licence withdrawal could occur if the operation of site facilities were below specified norms. Therefore, it was imperative nuclear system failures of any kind, be minimized and preferably prevented. With the benefit of 20/20 hindsight, it can now be said nuclear system failures were minimum: limited to a few that could have been considered major. This outstanding record must be attributed, in large part, to operator vigilance, good maintenance and design. WNRE was, and still is, blessed with highly skilled trades people, dedicated to performing good work and having an inherent consciousness towards safety.

As stated earlier, Buildings 412 and 408 were home to most of the site trades. Building 412 housed a machine shop, welding shop, inspection shop, sheet metal shop, instrument shop, tool crib, electrical shop and offices for engineering staff, supervisory staff, foremen, and lead hands. Building 408 housed a carpentry shop, automotive garage and offices for its supervisory staff, foremen and lead hands. In addition, "Stores Services", a branch of the Finance Division, shared office and storage space in this building.

Ground Maintenance and Mechanics Group – Roy Louden, Henry Wojciechowski, Mark Kaltenberger, Earl King, Vern Pommer, John Sauer, Brent Donnelly, Ted Faryon, Bill Sitar, Eric Ayres, Norman Bruneau, John Garbolinski, Bob Kost, Gerry Sachvie, Les Carlson

The machinists spent their time working in one area, as did the carpenters, tool crib attendants and auto mechanics. The welders spent some time in the welding shop, especially when they were constructing new experimental facilities, but spent the remainder of their time performing field welds wherever they were needed on site. It was most impressive watching them weld zirconium tubes together through a window of an inert glove box using the tungsten inert gas (TIG) process. When the weld was completed, it almost look machined it was so immaculate. Possibly not every welder on site could perform such outstanding work, but there were many that could.

The remainder of the trades spent little time in their home building. Depending on their specific work schedule they could be found anywhere on the site.

During scheduled reactor shutdowns, many trades would be involved in doing routine reactor maintenance, removing obsolete experimental facilities and installing new ones. Reactor instruments would be swapped, electrical circuitry would be checked

Fork Lift Training – Dennis Sol, Ben Kollinger, Dennis Graham, Doug Blais, Don Soluk, Bob Bruneau, Terry Kuhn, Gaylord Newman

and repaired, and all rooms unsafe for entry during reactor operations would be inspected, and if needed, would be serviced and made safe for ongoing operations.

Whiteshell Trades – 2012: Len Witoski (1), Dan Gagnon (2), Dean Antymis (3), Adrian Geary (4), Jack Bonekamp (5), Wayne Shewchuk (6), Rob Thomson (7), Grant Pachkowsky (8), Kelly Tymko (9), Cindy Litke (10), Blaine Grabowski (11), Ed Lowen (12), Edna Renard (13), Kevin Zieske (14), Bruce Murray (15), Sean Daymond (16), Darlene Keith (17), Jon Coss (18), Mike Veilleux (19), Brian Dyck (20), Kevin Clarke (21), Tom Kearns (22), Gavin Specaluk (23), Kevin Jackson (24), Wayne Leonard (25), Richard Abraham (26), Bill Lavallee (27), Ken Lodge (28), Larry Stelko (29), Dan Bannish (30), Phil Clarke (31), Ken Bilkoski (32), Jeff Bukoski (33), Eric Prokopchuk (34), Kirk Haugen (35), Quinn Dykstra (36), Nathan Dugard (37), Pete Kuzminski (38), Ken Urbanski (39), Greg Clark (40), Clint Veilleux (41), Bryan Koroscil (42), Bill Donald (43), Rod Henderson (44), Tim Breton (45), Derrick Zelinsky (46), Gary Rollins (47), Siggi Schuhmann (48), Chris Gulbinski (49), Dave Fillion (50), Tony Melendez (51), Dan Laliberte (52), Tim Barron (53), Kevin Lesosky (54), Garry Wood (55), Cory Dyck (56), Corey Abercrombie (57), Brian Favreau (58)

With the restart of the reactor, usually a fixed group of trades people was assigned to consolidate, decontaminate and classify items that could be reused or require disposal. Items classified as requiring disposal went through an established procedure of decontamination, classification, packaging and waste storage. Items classified as reusable also went through a decontamination and classification procedure, plus repair if needed, and were placed in storage for reuse. Performing this work in the reactor building meant that any radioactive contamination would remain in the building and was not transferred to other buildings or along roads and pathways between buildings. A similar maintenance philosophy was applied to the hot cells, were irradiated fuel and other radioactive items were examined and to other buildings located in the active area that could generate or receive radioactive products.

Powerhouse

The plant powerhouse, Building 911, was also part of the Maintenance Branch. It provided 200°C hot water to heat most of the buildings on site. The fuel used to provide the heat was Bunker "C" oil, trucked to the site. This was the first powerhouse built in Western Canada to use "hot water" for building heat and many held serious reservations it would work as predicted. However, it turned out to be an excellent system, except for the fuel. Bunker "C" is heavy oil and would congeal at winter temperatures. Trace heating at the burners, along the supply lines and the fuel storage tanks maintained the oil at an appropriate temperature for it to flow freely. However, the burning of bunker "C" was not a clean business. Periodically, the furnace required cleaning and this turned out to be a difficult and time-consuming task. So, a few years later the bunker "C" was replaced by light oil, similar to diesel fuel. This light fuel burned cleaner; cleaning the furnace was easier and no trace heating was required.

In about 1971, a decision was made to add a third circuit (called the C-circuit) to the WR-1 reactor. It was also decided to use the heat available in this circuit to supplement the plant site heating requirements. The maximum amount of heat the C-circuit produced while running at full reactor power approached 15 MWt. The plant heating requirements during the winter were high enough to use a significant portion of C-circuit's power output. During reactor operation, the powerhouse remained on warm-standby, allowing it to quickly takeover any heating duties in case power from the reactor became unavailable. During the ten-year period this system operated, energy savings were significant and the system operated without a hitch.

The winter of 1984 was a record breaker. Temperatures of -45°C were reached for a night or two. WR-1 reactor was operating at full power and C-circuit was delivering its maximum power. The powerhouse discharge hot water temperature recorder was showing a normal straight line at 200°C, until the early hours of these cold mornings, when this line slowly dropped about 2°C and stayed there until the late morning at which time it went back to the original value. This meant the site heating requirements were just being met by the maximum power output from C-circuit.

Some Powerhouse Personnel
Don Campbell
Harley Davidson
Everett Dobbin
D. Dubasov
K. Dykstra
Al Futcher
Marty Jacobs
Wilfred Keith
Ron MacLean
Bob Randell
John Rankin
R. Ruvmar

Surplus heat from C-circuit went to the river. Thought was given to how it might be better utilized and possibly commercialized. Greenhouse operations were one prospect, requiring heat from about January to early spring. Many greenhouse owners were approached and considerable interest was generated throughout the province and other parts of Canada, but in the end, no one would take the risk.

The powerhouse operators were also responsible for start-up, operation and shutdown of the chillers that provided cooled air to selected areas of the WR-1 reactor building. This cooling was important for the instrumentation and control systems. This normally involved a spring start-up and a fall shutdown with performance monitoring during operation. This chiller also provided summer cooling to other buildings on site.

The process water/domestic water/firewater pump house was located adjacent to the Winnipeg River. Domestic water was purified and distributed to most building for showers, washrooms and other domestic uses. Process water was distributed primarily to the WR-1 reactor to remove the reactor-generated heat from the organic coolant circuits and in-reactor experimental loops. Firewater was distributed to all on-site hydrants and to most buildings for use in firefighting. Diesel-driven back-up pumps were provided in the event of a pump house power failure.

Purchasing Branch

As with any large organization, AECL had a dedicated group of staff to support the operation of the corporation. One such group was the Purchasing Branch, probably one of the oldest groups associated with WL.

The WL Purchasing Branch actually began in 1963, as an office on St. Mary's Avenue in downtown Winnipeg. The branch moved to the WL site in August 1963, becoming one of the first occupants of the recently completed Building #402.

The first and longest serving manager of the group was Max Allan, serving from 1963 to 1981. The original branch had 3 senior purchasing agents, plus as many as 10 support staff. This small group was responsible for all purchases on a new site, with a newly assembled group of employees,

Max Allan

building a new nuclear reactor and the associated infrastructure. The times were both exciting and rapidly changing.

Al Peterson

Darlene Beeskau

Robert Tomchuk

Pat Porth

Shirley Benson

Ted Deering

The Purchasing Branch moved to various locations at the WL site over the years. The branch size has also varied to manage changing regulatory requirements and criteria for suppliers to sell to the nuclear industry. Many employees have been part of the branch, including people like Bud Kelly, Harold Peterson, Bob Tomchuk, Bernie Bjornson, Joan Ticknor, Pat Porth, Sandy Campbell, Dorthy Wilken, Carol Lee Augustine, Maureen Miller and Lynn Grant. One of the longest serving branch members was Darlene Beeskau. She joined the branch at the Winnipeg office in 1963. She became the first woman to be promoted to purchasing agent in 1975. Darlene retired from the branch in 1992.

The 2013 branch, now called Procurement, managed by Brian Bruneau, continues to play a vital role in the operation of the WL site. All products and services needed by the corporation go through this small group. Virtually every employee on site knows at least one branch member by name and works with them regularly to keep the site operating effectively.

Stores

Like the Purchasing Branch, Stores had and continues to have a dedicated group of staff to support the operation of WL, the URL site, and, in the earlier years, Kelsey House, the Pinawa Hospital and many small satellite sites such as Cigar Lake, Saskatchewan. The staff managed shipping and receiving tasks for all goods entering and leaving the site. Staff was required to be trained in classifying and handling of dangerous goods, including flammables, compressed gases, poisonous substances, explosives, and of course, radioactive shipments. They were also responsible for obtaining custom documentation for all shipments leaving the country.

Lloyd Dreger

Once items were received in Building 408, they were shipped to satellite stores locations in Buildings 300, 412, 420, 100, 415, 409, and 416. Stores departments also manned tool cribs in Buildings 412, 300, and 100. All tools were identified and handed out to the workers as required. Spare parts for older equipment used onsite were also stored in Building 412. In the late 1990's, the Stores Department also assumed responsibility for site mail delivery.

As the WL operations moved into decommissioning and equipment became surplus, the Stores Department broadened their mandate and began to manage equipment disposal. Before any piece of equipment could be sold, the Surplus Asset Maintainer needed to confirm that no other department on site needed the item. If it was not needed at WL, it was offered to CRL. If it was still deemed as surplus, the item was sold, initially through Crown Assets of Canada and, in later years, through direct sale to the general public. Surplus items were sold through a liquidator, using a public auction process, and a bid sale process. These sales took place twice a year.

During the 50th Anniversary year of WL in 2013, Stores, based in Building #408, was managed by George Gibson until October, followed by Randy Worona.

> **Some Stores Staff over the Years**
> Pat Williamson: Manager - B408
> Jim Kines: Traffic Coordinator Supervisor - B408
> Barney Chapman: Typewriter repair; Surplus officer - B408
> Joan Ticknor: Clerk - B408
> Cecile Schmidt: Secretary - B408
> Cas Szajewski: Inventory Control - B408
> Rosalie Lofstrom: Inventory Control
> Bob Lussier: Tool Crib - B412
> Bob Briercliffe - Tool Crib - B412
> Elmer Smith: Supervisor - B412
> Jim Rogocki: Counterman - B412
> Ed Komadowski: - Counterman - B300
> Wayne Wasylenko: Tool Crib – B300
> George Kowalchuk: Spare Parts - B412
> Lloyd Dreger: Supervisor - B408
> Chris Balness: Counterman - B415
> Ken Bjornson: Stores Keeper - B408
> Dave Germain: Counterman – B408
> Mike Furnish: Counterman – B412
> Don Wojcik: Warehouse Worker – B408
> Randy Mamrocha: Materials Logistics Lead Hand – B408

Pat Williamson

S. Reimer and Bob Lussier

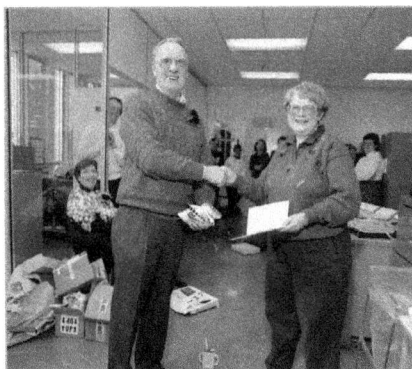

Jim Kines and Rosalie Lofstrom

Barney Chapman

Unions

Most, if not all of the trades, became unionized shortly after WNRE site operations began and later even went on strike on a few occasions. Other site personnel such as scientists, engineers, technologists, technicians, clerical, secretarial, etc., were not unionized initially. The unions representing the trades at WNRE in the early years were:

The International Association of Machinists and Aerospace Workers (IAM&AW), local 608, Certified in 1965: In the beginning, any trade person holding a ticket from a certified trade became a member of this union. They consisted of electricians, tool and die makers, machinists, control mechanics, instrument mechanics, welders, pipefitters, plumbers, auto mechanics, millwrights, stores keepers, drivers, labourers, RIS operators working in the laundry building and possibly others. A sub-group of trades from the local Government District of Pinawa (LGD) were also members of this union but they negotiated directly with the LGD. In the early 70's the pipefitters, plumbers and welders broke away from the IAM&AW because of pay level concerns and joined the UA.

Canadian Union of Public Employees (CUPE), Local 938, certified in 1965: The WR-1 reactor operators selected CUPE because Manitoba Hydro, Winnipeg Hydro and Ontario. Hydro operators already belonged to this union and it was felt they performed similar duties. The WR-1 operators remained members of this union until decertification in 1981.

United Association of Plumbers, Pipefitters and Welders (UA), local 254, certified in 1971: UA represented the pipefitters, plumbers and welders after they broke away from the IAM&AW in the early 1970s. UA represents approximately 340,000 plumbers, pipefitters, sprinkler fitters, service technicians and welders in local unions across North America. The UA provides five-year apprenticeship programs, extensive journeyman training, a comprehensive, five-year instructor training program, and numerous certification programs.

International Association of Fire Fighters (IAFF): IAFF represents the Whiteshell firefighters and were certified in 1971. Members of this union were considered to be an essential service and were not allowed to strike. The IAFF represents more than 300,000 full-time professional fire fighters and paramedics in more than 3,100 affiliates.

United Steelworkers of America (USWA): USWA represented labour and janitors who worked in the active area and were required to deal with and handle contaminated products. Contract labour was never part of this union nor did it work in the active area.

In the late 1980s, scientists and researchers formed an association to compare site similarities and disparities with those of equivalent professional background across Canada and make their findings available to the company. This was a first step towards unionization. The Professional Institute of the Public Service of Canada (PIPSC) was certified to represent the scientists and engineers at AECL Whiteshell laboratories on Feb. 14, 1996. This group is formally designated as the Whiteshell Professional Employees Group (WPEG).

PIPSC was also certified to represent technicians and technologists on June 24, 2011. This group is referred to as the Whiteshell Technical Employees Group (WTEG). They are the latest group of Whiteshell employees to be unionized. In a PIPSC news release dated Oct. 22, 2012, the following comment was made – "Having remained non-unionized for decades, Technicians and Technologists joined their AECL colleagues with a solid collective agreement. I'm proud that our union could assist at their time of greatest need", said Gary Corbett, President. This release was made after a strike against AECL was narrowly averted about a week earlier.

The *Canadian Nuclear Workers Council (CNWC)*, founded in 1993 is an umbrella organization of Unions representing workers in all sectors of the Canadian nuclear industry. CNWC activities are focused ensuring that the interests and perspectives of nuclear workers are heard by decision-makers and strengthening the collective role of nuclear workers as a partner in their industry. Whiteshell staff has been part of the CNWC since the mid-1990s.

Labour Relations

Labour relations between the company and unions were generally considered as good. Issues did arise, but negotiations usually resulted in agreements. However, not all negotiations were successful. Based on an article in the Whiteshell Gazette, dated, December 23, 1965, local 608 of the IAM&AW voted to take strike action against AECL if further negotiations fail to bring about an agreement. Differences between the union and AECL revolved around wages and benefits. A Federal Conciliation board tried to resolve these differences but neither side were in agreement with the board's recommendations. Both sides were near agreement on the wage package but not the benefits package. The article reported the outstanding issues remaining were: a first in, last out clause; company payment of 50% of the best Manitoba medical plan (instead of the lowest); a group insurance scheme; annual leave; transportation of employees to and from work; and granting of some degree of union security at WNRE.

The union issued a strike deadline of January 10, 1966. AECL offered the union a shorter contract term of 12 months. The union accepted the offer and a strike was averted. The union's other demands were dropped. The strike was averted in a large measure because Dr. Mooradian agreed to seek improved labour relations before the contract ended. AECL and union representatives agreed to study the issues of medical plans as it applied to all AECL trades personnel across Canada. Solutions were to be in place for next contract renewal in 1967.

The first strike at WNRE occurred in 1966. It was the first strike ever against AECL. The Whiteshell Gazette

IAFF 1987 Contract Signing: Don Trudeau, Jack Hildebrand, Bill Laird, John Stefaniuk, Ralph Green, Dave Beeching, Gib Drynan, and Bud Mager

reported the following: "Negotiations between AECL and local 938 of the Canadian Union of Public Employees (CUPE) recently ground to a halt, resulting in the strike of 28 members of the local and the picketing of WNRE after about 8 months of negotiations. AECL had offered increases matching those accepted by Chalk River and Ottawa unions but these were not accepted by CUPE as an existing differential between WNRE and the Ontario site was not eliminated. CUPE contends that this is an area of high living costs and that a wage structure based on low-area wages is not acceptable". There were other issues as well, as described by Lloyd Rattai with assistance from Larry Gauthier and Al Nelson.

"In the summer of 1964 a group of operations personnel transferred from Chalk River Ontario, to Pinawa Manitoba, to add to the experience required to commission and operate the new WR-1 reactor at the newly established Whiteshell Laboratories. WR-1 reactor was in the early stages of construction.

The eleven persons to make the trip from Chalk River with their families were (Foreman) Bennie Richter (Lead hands) Vinny McCarthy, Wilf Campbell, Roy Barnsdale, Dick Meeker, Mickey Donnelly, (First Operators) Larry Gauthier, Harry Backer, Al Nelson, Mike Berry and Ron Wiggins. Later that summer additional personnel were hired from the region. The boys from the West were Thor Borgford, Gerry Walters, Ben Kollinger, Gord Robinson, Jake Kruger, Jack Blacher, Gilles Marion, Ken Gray, Al Hawks, Lloyd Rattai, Albert Palson, Don Terschman, and Louie Bruneau. In

1965, a group of five technology graduates from RRCC were hired into operations. These were Fred Legiehn, Ron Fiecho, Bill Zuk, Ron Gurkie and Don Hatch.

During the hiring process, some new operators were led to believe that they would be classified as staff instead of prevailing rate. This did not happen. The significance of this classification change to staff would have been a difference of several important benefits such as increased vacations, sick leave and an improved pension plan. When the time came in 1965 to expect a raise in pay, the operators who transferred from Chalk River were told AECL had determined that Pinawa was a lower cost of living area than Chalk River. A survey of the cost of living comparisons proved the opposite was true. To revert to a lower rate of pay, lower than the comparative position in Ontario was not acceptable. It became clear after some discussions that we would have to unionize to be able to negotiate a fair condition of employment. In 1965, the operators were certified as the bargaining unit for the operators with the Canadian Union of Public Employees (CUPE 938). This union was a good fit as they represented the Hydro operators in Manitoba and Ontario.

The negotiating committee consisted of Roy Barnsdale, President of the Union, Larry Gauthier, Al Nelson and Lloyd Rattai. After several months of frustrating meetings, the membership was asked for a mandate to take strike action against AECL, the first in AECL history. The strike vote was supervised by Keith Norton and Bud Henderson, staff representatives from CUPE.

On February 15, 1966, WR-1 was shut down in anticipation of a work stoppage. On the morning of February 16, the picket line went up at the junction of Highway 211 and the plant road. The temperature was -33ºF (-36.1ºC). The sight of the determined individuals slowly walking across the intersection, could remind one of refugees in a war movie. The kind of winter clothing available now, was not readily available at the time and certainly not part of the normal clothing of the men involved. On February 17, 1966, Edward Schreyer, MP for Springfield, asked Jean Luc Pepin, the Minister responsible for AECL what he was doing to resolve the strike in Pinawa. The Minister would take it under advisement. On the 18th, the temperature dropped to a record 56ºF below 0 (-48.9ºC).

The strike dragged on with no break in negotiations. Significant financial support started to arrive from other CUPE locals for strike pay. Many families were getting very low on funds. In late February, Ed Schreyer flew out to our union meeting in Pinawa to get our concerns first hand. The next day, back in Ottawa, he asked Minister Pepin if his government was interfering and limiting AECL in their negotiations in Pinawa. The media picked up the story and the Whiteshell Manager was called to Ottawa. A meeting in Winnipeg with a mediator was arranged for March 8, 1966. Pinawa was isolated from the outside world on March 4th by a blizzard with record winds and snowfall. On March 8th, the negotiation committee was able to get to Winnipeg under very difficult conditions to negotiate a tentative agreement with AECL. On March 10th, the members of CUPE local 938 voted to accept the offer and return to work. Most of the issues were resolved some time later when AECL switched operating personnel to a staff classification."

The reactor operators never went on strike again. They were decertified in 1981. The operators are now represented by the Professional Institute of the Public Service (PIPS), becoming certified in 1996.

AECL's second strike at Whiteshell occurred in the fall of 1971. Forty members of the Plumbers and Pipefitters Union who were part of the Winnipeg Plumber's Union Local 254 went on strike on September 21. The reactor was shut down as a precaution as 200 other union members at the site honoured the picket lines. The key issue was a demand for reclassification by the union to a higher trade classification to reflect the type of work being done at Whiteshell. The strike ended in early 1972.

A third strike against AECL was initiated by UA, local 254, in 1973. This was an eight-week strike with both the Whiteshell and Chalk River laboratories being involved. According to a news clipping dated Nov. 20, 1973, the following was written; "A joint press release issued by AECL and the Atomic Energy Allied Council early last week has announced the major points of agreement which led to the settlement of an eight-week strike at the Whiteshell Nuclear Research Establishment and at the Chalk River Nuclear Laboratories." It then listed the benefits. This was the last strike that the WL endured, although union activity continued and even expanded to include Scientists, professionals, technologists and technicians.

Chapter 4
Completing the Mission

One of AECL's major achievements was the preparation and public defense of a ten-volume Environmental Impact Statement (EIS) on the concept of deep geological disposal of Canada's nuclear fuel waste in 1994. This project was primarily based on work conducted at the Underground Research Laboratory (URL).

In 1978 the governments of Canada and Ontario established the Nuclear Fuel Waste Management Program (NFWMP). Responsibility for research and development on disposal in a deep underground repository in intrusive igneous rock was allocated to AECL. Ontario Hydro, Natural Resources Canada, Environment Canada, universities, and consultants in the private sector also contributed to the NFWMP.

AECL believed that human health and the natural environment must be protected in any disposal concept; the burden placed on future generations must be minimized, there must be scope for public involvement during all stages of concept implementation; and the concept must be compatible with the geographical features and economic factors in Canada. AECL described its concept as a method for geological disposal of nuclear fuel wastes in which:

- the waste form is either used CANDU fuel or solidified high-level wastes from reprocessing;

- the waste form is sealed in a container designed to last at least 500 years;

- containers of waste are placed in rooms in a disposal vault or in boreholes drilled from the rooms and surrounded by a buffer material;

- the disposal rooms are between 500 and 1000 metres below the surface;

- the geological medium is plutonic rock of the Canadian Shield;

- each room is sealed with backfill and other vault seals; and

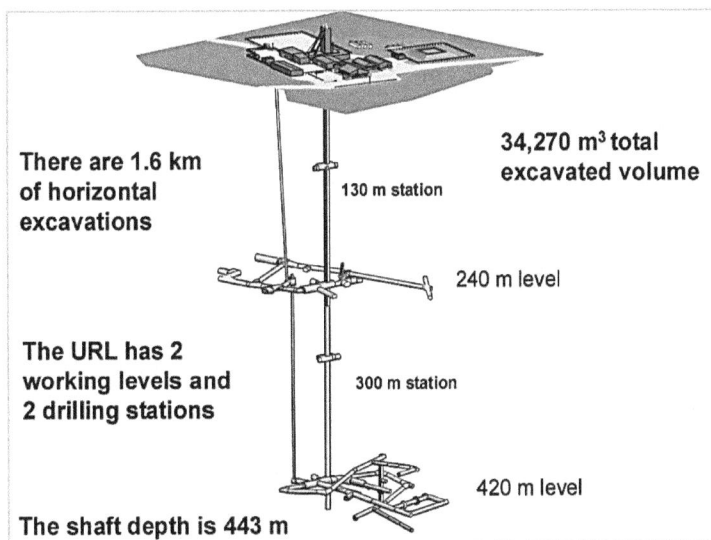

There are 1.6 km of horizontal excavations

130 m station

34,270 m³ total excavated volume

240 m level

The URL has 2 working levels and 2 drilling stations

300 m station

420 m level

The shaft depth is 443 m

Excavation at URL

- all tunnels, shafts and exploration boreholes are sealed so that the facility would be passively safe; safety would not depend on institutional controls.

AECL consulted broadly to help ensure that their concept was technically sound and represented a generally acceptable disposal strategy. Many groups in Canada had opportunities to comment on the concept and on the NFWMP, including government departments and agencies, scientists, engineers, sociologists, ethicists, and members of the public.

Underground Research Laboratory

The Underground Research Laboratory (URL) shaft selection and surface facility construction began in 1982. The URL was situated on the western edge of the Precambrian Shield, approximately 20 km east of WL. AECL constructed the facility to provide a representative geological setting for conducting research and development activities in support of the Canadian NFWMP.

The objective for the URL was characterize of the rock mass, groundwater flow systems and groundwater chemistry of the geologic environment, and to complete studies of the engineered components of the repository sealing system.

Phases of the URL

Siting Phase

The URL siting phase started in 1978. A small set of screening criteria was established for selecting a site. The site had to be larger than 1 km^2, be predominantly outcrop, and be undisturbed by previous excavations. The site had to be within, but not close to, well defined hydrologic boundaries. The site had to be accessible, near power, near AECL's Whiteshell Laboratories and available for lease. Eight potential sites were identified.

Site Evaluation Phase

The site evaluation phase was carried out between 1980 and 1983. The objective was to develop an approach for designing and constructing a repository in granite. This phase involved surface mapping, airborne and ground geophysical surveys, surface water and meteorological data collection, and the drilling of boreholes for piezometric measurements. The detailed characterization revealed three low-dipping fracture zones that controlled the groundwater movement and groundwater chemistry within the rock mass. The location of the shaft was in a region with moderate fracture zone permeability to permit future underground experiments. Based on the experience gained at the URL, an approach to underground characterization for a deep geologic repository was developed.

Vern Steiner

Construction Phase

Shaft collar excavation and construction of the surface facilities were completed in 1983. Excavation of the shaft to a depth of 255 m occurred in 1984. A loop of horizontal excavations on the 240 m level and the ventilation shaft were completed by 1987. The main shaft was extended to a depth of 443 m in 1988, followed by the excavation of the 420 m level and the ventilation shaft over the next three years. The primary objective was always to provide a safe and efficient underground research facility. The guiding principle was to maximize the benefit to the research program in order to best achieve the objectives set out for the URL.

AECL's Underground Research Laboratory

Operating Phase

The operating phase experiments were developed in 1989 and underwent peer review by a panel of leading Canadian scientists. The program included seven major experiments and two experimental programs:

- Solute Transport in Highly Fractured Rock Experiment
- Solute Transport in Moderately Fractured Rock Experiment
- Grouting Experiment
- Buffer/Container Experiment
- Shaft Sealing Experiment
- Mine-by Experiment
- Multi-Component Experiment
- In situ Stress Program
- URL Characterization Program

The experimental program began in 1990. Experiments preformed prior to 1997 focused on demonstrating the safety and feasibility of AECL's concept. Work after 1997 was directed towards addressing identified gaps in technologies required to construct and license a deep geologic repository.

Art Holloway

An example of the post-1997 work was the Engineering Design of Repository Sealing Systems (ENDRES) project. The ENDRES project was designed to identify and address gaps in sealing technology.

The overall objective was to develop engineering tools to optimize the design of repository sealing systems. These tools were categorized into three groups: numerical models; instrumentation; and physical tests. This project attempted to link together the results of a number of in situ experiments. A number of projects were also performed to address many of these gaps.

Work conducted at the URL contributed greatly to the Canadian NFWMP and the EIS submitted to the federal government in 1994. The work also supported waste management programs for other countries that did not have a world class facility like the URL.

Government Response

After extensive public consultations, the federal government concluded that AECL's disposal plan could only be acceptable if it

- had broad public support;
- was safe from both a technical and a social perspective;
- was developed within a sound ethical and social assessment framework;
- had the support of Aboriginal people;
- was selected after a comparison with the risks, costs and benefits of other options; and
- was advanced by a stable and trustworthy proponent and overseen by a trustworthy regulator.

A concept for managing nuclear fuel wastes must also be judged to

- demonstrate robustness in meeting the regulatory requirements;
- be based on thorough and participatory scenario analyses;
- use realistic data, modelling and natural analogues;
- incorporate sound science and good practices;
- demonstrate flexibility;
- demonstrate that implementation is feasible; and
- integrate peer review and international expertise.

After applying these criteria, the government's Evaluation Panel concluded that from a technical perspective, the safety of concept was adequately demonstrated. However, AECL had not demonstrated that the concept for deep geological disposal had broad public support. Therefore, the concept did not have the required level of acceptability to be adopted as Canada's approach for managing nuclear fuel wastes.

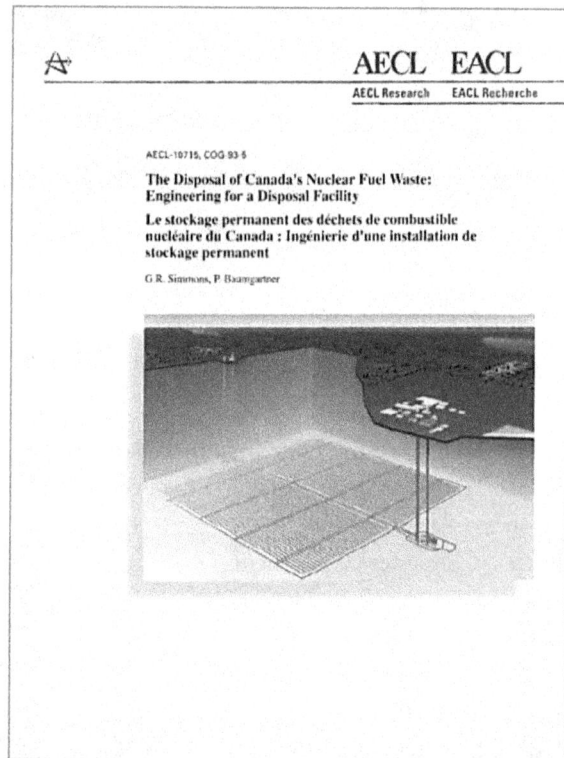

URL Staff: Doug Peters, Colin Allan, Glen Snider, Dave Woodcock, Glen Karklin, Ben Delannoy, Hugh Spinney Ray Fillion, Bob Hampshire, Dick Winchester, Vern Steiner, Shawn Keith, John Wedgewood, Peter Roach, Larry Rolleston, Cliff Kohle, Ken Dormuth, Gary Wallace, Gary Simmons

A number of additional steps would be required to secure broad public support, including:

- issuing a policy statement on managing nuclear fuel wastes;
- initiating an Aboriginal participation process;
- creating a nuclear fuel waste management agency (NFWMA);
- conducting a public review of regulatory documents using a more effective consultation process;
- developing a comprehensive public participation plan;
- developing an ethical and social assessment framework; and
- developing and comparing options for managing nuclear fuel wastes.

URL hosting international guest – Ray Sochaski, Gary Simmons, Guest, Cliff Davison

The Government of Canada accepted the recommendations of the Panel. The Nuclear Fuel Waste Act was passed and came into force in November 2002. The act required the nuclear energy corporations in Canada to form a waste management organization. As directed, the corporations formed the Nuclear Waste Management Organization (NWMO). The act also required the establishment of a segregated fund for nuclear fuel waste management in Canada, with funding coming from all of the nuclear utilities and AECL. The NWMO's mandate was to study options for the long-term management of nuclear fuel waste and recommend a preferred option. The options included deep geologic disposal, long-term storage at reactor sites, and long-term centralized storage above or below ground. The NWMO continues its work today, as per the requirements of the Nuclear Fuel Waste Act and oversight by the Minister of Natural Resources.

Glen Snider, Dwayne Onagi, Richard Suski, Don Daymond, Glen Karklin, Shawn Keith, Vern Steiner, Jason Martino, Des Johnson, John Wedgewood

Chapter 5
Commercializing Science

AECL has a rich history in nuclear Science, Technology and Innovation (ST&I), from commercializing the CANDU power system around the world, advancing fuel technology and nuclear safety, to protecting human health through nuclear medicine and cancer therapy technology. The research at Whiteshell has been a part of this proud history. Whiteshell research has led to new products and business opportunities, including reactor design improvements, analytical equipment, and waste management techniques. Whiteshell staff has also contributed to the commercialization of selected industrial applications for radiation, nuclear safeguards technologies and environmental research.

SLOWPOKE Reactors

The "SLOWPOKE" or Safe Low-Power Critical Experiment is a low-energy, pool-type nuclear research reactor designed by AECL in the late 1960s. This type of reactor provided a higher neutron flux than other small commercial accelerators, while avoiding the complexity and high operating costs of existing nuclear reactors. The reactor utilized a very small quantity of uranium fuel, and was surrounded by a beryllium reflector that scattered neutrons back into the core, maintaining the fission chain reaction. By adding thicker pieces of beryllium to this reflector region it was possible to extend the life the original fuelled core to several decades.

A SLOWPOKE research reactor was conceived at Whiteshell in 1967. In 1970, a prototype unit was designed and built at CRL. This prototype was moved to the University of Toronto in 1971. It had one sample site in the beryllium reflector and operated at a power level of 5 kW. In 1973 the power was increased to 20 kW and the period of unattended operation was increased from 4 hours to 18 hours.

The first commercial version of the SLOWPOKE was started up in 1971 at AECL's Commercial Products Division in Ottawa. In 1976 an improved commercial design, named SLOWPOKE-2, was installed at the University of Toronto. The commercial model had five irradiation sites in the beryllium reflector and five sites stationed outside the reflector.

SDR Location in B100

Between 1976 and 1984, seven SLOWPOKE-2 reactors with Highly Enriched Uranium (HEU) fuel were commissioned in six Canadian cities and in Kingston, Jamaica. In 1985 the first low-enriched Uranium (LEU) fuelled SLOWPOKE-2 reactor was commissioned at the Royal Military College of Canada (RMC) in Kingston, Ontario. Since then several units have been converted to LEU, as part of a U.S.-led program to reduce the amount of HEU in use around the world in civilian reactors.

In the early 1980s AECL designed and built a scaled-up version called SLOWPOKE-3 for district heating at Whiteshell. The Slowpoke Demonstration Reactor (SDR) construction began in 1987. The pool type water-cooled reactor was designed to demonstrate the viability of the reactor to provide energy for small communities or comparably sized industrial establishments. The reactor was a 2 MW, unpressurized pool-type reactor.

The reactor core, hot riser duct and primary heat exchangers were installed in a water-filled pool, inside a stainless steel-lined, below-ground concrete vault. The core consisted of 4 fuel bundles containing 113 kg of 4.9% enriched uranium oxide, surrounded by a beryllium reflector. The pool was 4.3 m in diameter by 9.8 m deep and contained 121,000 L of water.

Primary cooling was by natural circulation through two plate-type heat exchangers. Water was heated by the core to 90°C and directed by the duct to the two heat exchangers where the heat is transferred to the secondary system. The primary flow was cooled and flowed down the outer annulus of the pool back to the core. The secondary heat transport system was a pumped closed circuit loop, which removed heat from the primary system heat exchangers and transferred the heat through the secondary system heat exchanger to water from the Winnipeg River.

The SDR was equipped with two shutdown systems; a gadolinium nitrate liquid absorber safety system and a fast shutdown system incorporating four gravity-drop absorbers. The four absorber plates used for shutdown were also used for reactivity control. The central control rod was used to regulate the power and provide automatic control of coolant temperature. Temperature

SDR Reactor Pool

sensors in the primary coolant provided the information to control the reactor. Neutron instrumentation provided feedback signals for manual and automatic control of the reactor. Pool water was pumped through ion exchange columns to maintain water chemistry and to control corrosion. The ion exchange column was also designed to remove fission products from defective fuel, and gadolinium nitrate from the liquid absorber shutdown system.

The SDR operated for a total of 614 hours between 1989 and 1990, at which time the reactor was permanently shut down and steps were taken to place the reactor in a secure state in preparation for decommissioning.

The economics of a district-heating system based on SLOWPOKE-3 technology were initially estimated to be competitive with that of conventional fossil fuels for use in remote communities. However market interest in the SLOWPOKE heating system eventually dwindled due to the low price of natural gas. Currently, the high price of oil and natural gas has sparked renewed interest in the use of nuclear energy for district heating purposes.

Nuclear Battery

The Nuclear Battery was designed to generate electricity and/or high-grade steam heat. It was being pursued by AECL as a complementary, follow-up product to the SLOWPOKE-3.

The program originated in 1984 as a joint project with the Los Alamos National Laboratory (LANL) to develop a small nuclear power supply for unattended short-range radar stations in the new North Warning System (NWS). Although no nuclear reactor design can claim absolute inherent safety, reactors can be designed with safety attributes that will prevent the catastrophic release of radioactive fission products in severe accidents. Passive safety features that cannot be overridden by misguided intervention were of paramount importance throughout the Nuclear Battery program. The coated-particle fuel chosen for the Nuclear Battery embodied perhaps the ultimate in containment principles because each fuel particle was independently protected. The small particle size enabled its coating layers to withstand potential internal pressures and stresses. Indeed, even dispersion of the core for whatever reason would not negate the containment properties of the individually coated particles.

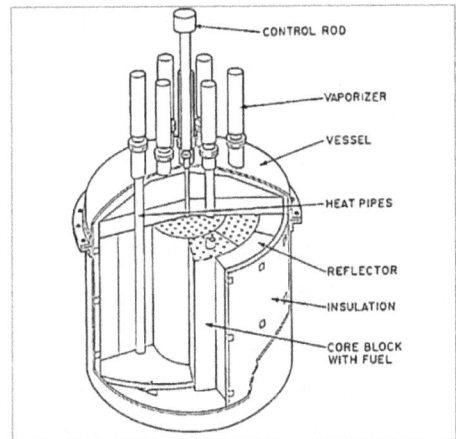

Nuclear Battery Reactor Core Module

The most outstanding safety feature of the fuel was its demonstrated ability to withstand extreme temperatures. No failures of the particle coatings would be induced by exposure to a temperature of 1600°C for up to 100 h, well below the peak fuel temperature in the Nuclear Battery of about 700°C with all four control rods fully withdrawn.

In hardware development programs at Whiteshell, a toluene organic Rankine cycle engine was commissioned and operated with a propane heat source. This demonstration-scale unit was operated for a total of 784 h, including 465 h in a single continuous run. It was then converted to an electrically heated configuration that permitted the determination of its conversion efficiency as a function of its operating state while providing data on toluene thermal degradation rates. Decomposition experiments concerning gamma radiolysis were conducted with static capsules and flowing-loop tests.

An experiment to study graphite oxidation under conditions that roughly simulate an air ingress accident in the Nuclear Battery was also performed. A small, electrically heated graphite block was consumed over a six-week period at temperatures progressively increased from 600 to 800°C, without observing a temperature runaway.

The project was cancelled when it became apparent that a full-power demonstration could not be completed in time to meet the demanding deployment schedule for the NWS application.

Hydrogen Recombiner

Research at Whiteshell included hydrogen combustion phenomena and mitigation technologies. One example was igniters, which could be used to burn off excess hydrogen. Another example was the demonstration of an alternative hydrogen combustor for use in vault cooler ducts where deuterium can accumulate.

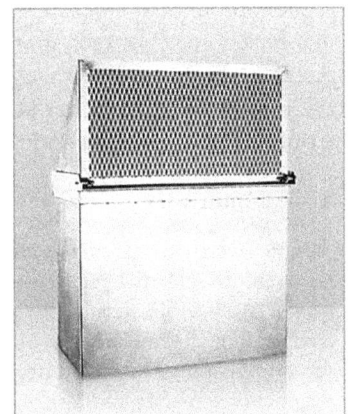

Controlling hydrogen in containment – recombiner technology.

The technology that got commercialized was AECL's Passive Autocatalytic Recombiner (PAR). PARs automatically activated without the need for external power or operator action, effectively preventing hydrogen buildup in the contaminated structure. Hydrogen is catalytically combined with oxygen in the air to form water vapour. An exothermic reaction occurs at the surface of the catalyst plates when hydrogen and oxygen are present in the atmosphere. The heat of the reaction, combined with the vertical arrangement and spacing of the catalyst plates promote natural air flow through the recombiner. Warm humid air and is exhausted through the top grating while fresh air and hydrogen are drawn through the bottom.

Industrial Applications of Radiation

The Radiation Applications Research Branch (RARB) was formed in 1985 to carry out research on industrial applications of ionizing radiation. Work at CRNL had led to the development of a high energy (10 MeV), high power industrial electron accelerator. It was believed that a significant commercial opportunity existed for developing new applications and sales of accelerators.

A key facility for RARB was the pilot-scale Electron Irradiation Facility built at the west end of building 402. The I-10/1 (10 MeV, 1 kW) electron accelerator provided the source of radiation and a conveyor system to moved product to the accelerator for treatment and back to the shipping area. The facility was useful for pilot-scale batches of material as well as experimental irradiations.

The RARB worked with industry to understand and solve the practical problems of using an accelerator to treat commercial products. A major focus was to study the use of electron processing to decrease the incidence of bacteria including salmonella, listeria and *E. coli* in

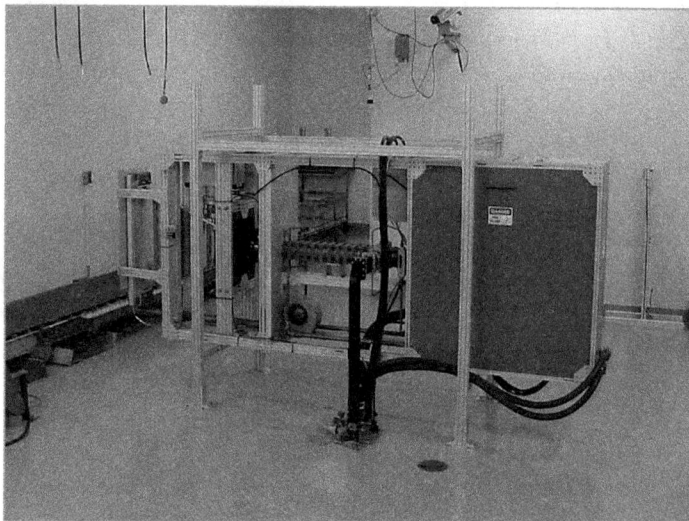

Acsion's 4 kW, 10 MeV accelerator purchased in 2002

common foods like chicken, hamburger and fish. A second thrust was to extend the range of products polymerized, crosslinked, grafted or broken down by irradiation. Early studies on polychlorinated biphenyls (PCB) and sewage sludge demonstrated technical feasibility but the applications were not economically feasible. Other key technical advances included electron beam curing of composite products, the production of viscose and the breakdown of heavy oil products.

Acsion Industries

Acsion Industries is a technology development company that was formed in 1998 to commercialize the AECL research on industrial applications for radiation. Acsion, through their various divisions and subsidiaries, focuses on the 'science of safety'. They serve public and private clients in the areas of energy, health and radiation safety, environmental protection and material science. Acsion has been engaged in several lines of business, including the production of medical isotopes, radiation treatment of products in the healthcare, agricultural and consumer products sectors, decommissioning of radioactive sites in Canada, providing training courses to make sure skilled workers are available with knowledge of radiation, and aircraft repair services at Acetek Composites. Acetek was formed in 2001 as a joint venture with Air Canada. Acsion purchased Acetek from Air Canada in 2004

Safeguards Instrumentation

In the mid-1980s, the International Atomic Energy Agency (IAEA) requested help to develop instrumentation to verify spent fuel discharged from nuclear reactors into spent fuel storage ponds. Three countries took up the challenge to develop an instrument: the United States, Japan and Canada. The Canadian effort was initially funded by the Atomic Energy Control Board (AECB) and Atomic Energy of Canada (AECL) and later fully funded by the AECB. The Canadian effort at Whiteshell focused on a modified night vision system that used an image intensifier with enhanced sensitivity to view the ultraviolet region of the Cerenkov spectrum. To permit operation in an illuminated fuel bay, a special filter was developed that had high transmission of light in the ultraviolet region and extremely high rejection of visible light.

The Canadian Safeguards Support Program (CSSP) was one of the first programs with an objective to assist the International Atomic Energy Agency (IAEA) by providing technical assistance and other resources to develop equipment to improve the effectiveness of international safeguards. The program comprised of five major activities: human resource assistance, training, systems studies, equipment development and support, and information technology.

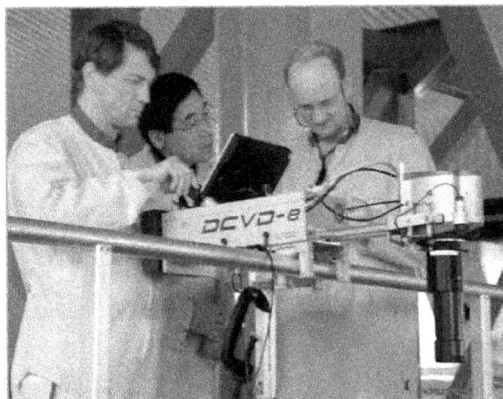

Dennis Chen and Brian Wilcox using DCVD instrument

Whiteshell scientists were the driving force behind the development of the Cerenkov Viewing Device (CVD) at AECL. The CVD was recognized by the IAEA as the instrument-of-choice for safeguards inspectors conducting fuel inventory verification at nuclear installations. In 1983, a team at Whiteshell developed the prototype, adapted from military night-vision technology. The CVD let inspectors quickly and non-invasively view the residual radioactivity in used nuclear fuel stored under water at reactor sites. This technology brought Canada international recognition for technological excellence and dedication to nuclear safeguards.

Channel Systems

Channel Systems is a company formed to commercialize nuclear safeguards technologies. They have grown into a global leader in these technologies. Channel Systems supplies the Digital Cerenkov Viewing Device (DCVD) for non-intrusive inspection and verification of used nuclear fuel.

The innovative features of the DCVD make it a valuable alternative to other verification measures. The real-time display and the capture of digital images for analysis and comparison make the instrument simple and highly effective. They also deliver high-quality training programs on the equipment and interpretation of the images gathered. Channel Systems have been lead instructors to the IAEA Spent Fuel Verification courses since 1989. Channel Systems has firsthand knowledge of the characteristics and regulations regarding nuclear fuel, its storage, and inspection. They also expanded their business to include custom solutions for measurement and automation needs, system designs, integrating high-performance, PC-based measurement, automation, data acquisition and control systems, and machine vision technologies.

Environmental Studies

One of the first steps in commissioning the WNRE site was to undertake a radiation survey in surrounding area. Routine environmental surveillance at began in 1962 to ensure worker safety, improve mitigation measures and to develop appropriate responses to unforeseen events. Routine surveillance carried out on samples of water, air, fish, plants and soil ensured that releases of radioactive materials were below specified limits and acted as checks on the models that were being developed. The doses to the public due to radioactivity in WL effluents have always been very small compared with background radiation. Aspects of this work continue today.

Steve Sheppard

Environmental research at Whiteshell focussed on the transfer of radionuclides from soil and water to plants and animals. Sampling was done across Canada to support several research programs.

Whiteshell was a world leader in environmental protection, using the many unique facilities at the laboratory to examine how radiation interacted with the environment.

ECOMatters

In 1998, ECOMatters Inc. was formed to focus on environmental research. Their first contract investigated the toxicity of metals released by copper and zinc smelters across Canada, a good adaptation from the previous 'nuclear' work.

Work came from governments and the private sector in Canada, France, Sweden, Japan, Finland and the United Kingdom. Research areas included nuclear, toxic metals and agricultural interests. Samples collected include fish from the arctic, deer stomachs from Manitoba, apples from Nova Scotia, soil from Sweden and manure from pigs in the Ottawa Valley. ECOMatters was also noted for writing computer models, including a series of models to estimate ammonia emissions from livestock in Canada.

ECOMatters continues to offer both nuclear and non-nuclear environmental consulting and research services. The Ecomatters' team specializes in detailed understanding of basic processes in terrestrial and aquatic environments, including the partitioning of contaminants and their uptake by plants and animals. Developing, applying and reviewing environmental computer models continue to be a major activity at the company.

Scientific publication also remains a cornerstone of the business. ECOMatters is home to the Editor-in-Chief of the Journal of Environmental Radioactivity. ECOMatters also houses the Head Offices of the Canadian Society of Soil Science and the Canadian Society of Agronomy.

Chapter 6
Decommissioning Whiteshell

AECL made a business decision in 1997 to transfer research programs to CRL and begin the process of decommissioning and closing Whiteshell. The Whiteshell Laboratories Decommissioning Project was funded under the Nuclear Legacy Liabilities Program (NLLP). The NLLP, established in 2006, was a program to manage Canada's nuclear legacy liabilities at AECL sites. The program was funded by the Government of Canada through Natural Resources Canada (NRCan).

A comprehensive environmental assessment to decommission Whiteshell was successfully completed in 2001. Decommissioning activities began under the first decommissioning licence issued in 2002 by the CNSC. The current licence, issued in 2008, is in effect until 2018.

AECL's Decommissioning Strategic Plan was developed to lay out the path to decommissioning all facilities and infrastructure at WL. The plan contained timelines, priorities, schedules and cost estimates. The plan also contains assumptions, decision schedules, risks and opportunities. The planned end-state of WL Decommissioning Plan was the complete decommissioning of the WL Site, transfer of wastes for disposal and release of lands.

The Approach

Whiteshell staff developed a waste management strategy which included waste handling. The plan was developed to support AECL's obligations to protect the health and safety of the public, our employees and the environment, and AECL's vision that we will minimize nuclear legacy obligations for future generations. It included an overview on projected waste volumes generated through the ongoing operations, decommissioning activities and planned remediation and retrieval of waste.

In general, the work involved the complete shutdown, decommissioning and decontamination, removal of all hazards, demolition and site restoration. A final radiological clearance survey of a building footprint would be conducted to verify and demonstrate that the remediated site met the final release criteria.

Demolition of Cafeteria

Waste Management Project

The Waste Management Project evaluated the current waste storage facilities and determined how to best reduce the current volume of historic waste, condition and re-package the waste, and ensure that it is stored safely. The concrete Medium-Level Waste (MLW) storage bunkers were assessed to ensure that they remained fit-for-purpose and remediated as required. Plans were developed to empty and decommission the standpipes, vertical concrete structures used to store MLW. Expansion of the WMA was also considered, including the dry-storage concrete canisters.

Shielded Modular Above-Ground Storage (SMAGS) – Building 923 (B923): The SMAGS was constructed in 2010, providing 4000 m³ of low-level radioactive waste storage. Prior to placement for storage in SMAGS, low-level radioactive waste (LLRW) is characterized, highly compacted and packaged in engineered containers. The building incorporates radiation shielding using 36 cm-thick pre-cast concrete components. The concrete foundation and a below-foundation membrane system provide containment barriers to the environment. One of the critical construction activities was a 22-hour, single-pour, concrete foundation.

Shielded Modular Above-Ground Storage Building

As a first use of SMAGs, waste generated during the Phase 1 decommissioning of the WR-1 Facility in the 1990's was retrieved, compacted and repackaged and either placed in the SMAGS building or shipped off-site for processing.

Soil Storage Compound (SSC): In order to provide safe, interim storage of low-level contaminated soil, a SSC was constructed at the WMA. The SSC is an open mound with environmental barriers designed to hold 2,000 m³ of low-level contaminated soil. The environmental barrier system is bentonite and high density polyethylene sheet.

Medical, Environmental and Dosimetry - Building 402 (B402): Building 402 housed the dosimetry services and the environmental monitoring program and laboratories. In its final years it housed commercial tenants such as Acsion Industries and Channel Systems.

Research and Development Complex - Building 300 (B300): Building 300 was the primary research laboratory at Whiteshell.

Environmental Barrier for SSC

Comprising seven construction stages and a floor area of approximately 17,000 m², Building 300 housed laboratories, offices, mechanical rooms and a high bay area for large-scale engineered experiments.

Active Liquid Waste Treatment Center (ALWTC) - Building 200 (B200): The ALWTC began operation in 1963. The effluents were transferred via underground lines to B200 prior to sampling, treatment and controlled release. The ALWTC was essentially an indoor tank farm. Pairs of tanks accept liquid waste from various sources on the WL site. While one of the pair of tanks was filling the other was being processed.

Building 200

The facility included a system that concentrated liquid waste originating from the Shielded Facilities (SF), the WMA and WR-1 in a resin-based solidification process. The concentrate was solidified and then transferred to the WMA for storage. There were a total of four sumps in the building that collected the various liquid wastes from the floor drains, leaks and tank overflows. The sumps collect and return the liquid waste to designated holding tanks. Safety systems included emergency lighting, remote and local process alarms (tank and sump high liquid levels), fire alarms, and radiation alarms.

Decontamination Center - Building 411 (B411): The Decontamination Centre (B411) was first commissioned in 1966, primarily as a laundry facility. There is a below-ground 2,200 L drain tank for interim storage of solvent-based mixed wastes arising from decontamination activities. There are also two holding tanks located in the basement that are used during the transfer of laundry and decontamination effluents. These effluents were eventually pumped to the ALWTC to be sampled, filtered and discharged, as appropriate, to the Winnipeg River. The safety systems in B411 consisted of emergency lighting, remote and local process alarms (for high liquid levels in tanks), and fire alarms.

Building 411

WR-1 Reactor Facility - Building 100 (B100): Phase 1 decommissioning of WR-1 and B100 was completed in 1995. Easily mobilized radioactivity (fuel, fluids, etc.) was removed from the facility. The main floor (600-level) and first sub-level (500-level) were decontaminated. Phase 1 work substantially reduced the potential hazards from the facility and reduced the monitoring and surveillance requirements for the deferment period. Detailed planning for Phase 2 of WR-1 decommissioning began in 2012. The original plan was to remove and package all activated and contaminated components, decontaminate the facility structure, demolish the building and remediate the site, meeting the final release criteria. Completion was planned for 2025.

Shielded Facilities (SF): The Shielded Facilities include the Hot Cell Facilities (HCF) and the Immobilized Fuel Test Facility (IFTF). The long-term strategy was the decommissioning, demolition and site remediation of the Shielded Facilities.

Strategy Going Forward

AECL's Decommissioning Strategic Plan laid out the path to full decommissioning of all facilities and infrastructure. The Plan was presented in five broad themes:

1. Decommissioning and environmental remediation, comprising all facilities, grounds and structures.

2. Waste management, including construction of facilities for handling, processing and storage of waste, plus the maintenance or remediation of the existing WMA.

3. Infrastructure modifications needed during decommissioning.

4. Maintaining support functions such as general site services, administration, and safety and licensing programs,

5. End states, including the requirements and processes to comply with the Decommissioning Licence and all regulations.

In September 2015, AECL contracted Canadian National Energy Alliance (CNEA: www.cnea.co) to take over Canadian Nuclear Laboratories and complete the decommissioning of the Whiteshell site. CNEA and CNL are now working with AECL to finalize the schedule for completing the decommissioning and the final end-state for the site, including the WR-1 reactor and the WMA. It is currently anticipated that decommissioning activities would be completed by 2026, with only long-term care and maintenance activities remaining for small areas at the site.

Dismantling B100

Barry Stefaniuk

Cleaning up the cesium pond

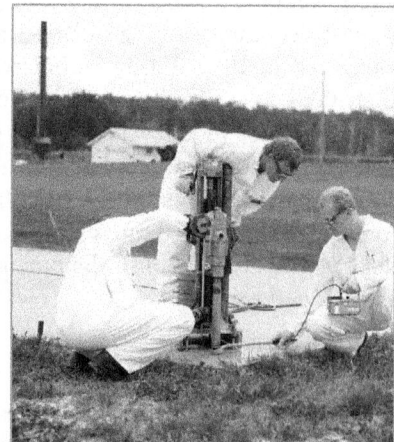
Wendal Schatkowsky and Jeff Hayter

Chapter 7
Life at Whiteshell

Whiteshell's contribution to nuclear science and engineering in Canada is indeed impressive. However, the Whiteshell story can not only be told through the thousands of peer reviewed publications, conference papers, presentations, and technical reports. You also have to look at the employees and their stories. Whiteshell has been the working home of thousands of scientists, engineers, technicians, trades peoples, and support staff. Do you remember your employee badge number?

Employees travelled from around the world to be part of the Whiteshell team. Some were here for only a few years. Many chose to have a career, working for AECL for decades. Second and even third generations are now working at Whiteshell. AECL provided over 33,000 years of full-time employment for Whiteshell staff from 1962 to 2015. The site began with 27 full time employees in 1962, growing to as high as 1060 in 1985. In addition, there have been thousands of casual, part-time, term and student employees that have been part of the Whiteshell story.

This final chapter tries to capture what it felt like to work at Whiteshell Laboratories through pictures. From the earliest days building the research site, to completing the many important research programs, to decommissioning of the site 50 years later, Whiteshell Laboratories has been an important part of Canadian nuclear family. We have made friends for life that cannot be described in a technical report.

Look at the pictures. Remember your friends and colleagues. Follow the newspaper headlines and selected publications. We hope you get a feel for what it was like to be part of the Whiteshell family.

To the outside world the most noticeable feature at Whiteshell was the WR-1 "stank", a combination emergency coolant tank and ventilation stack.

Starting the adventure

Watching the construction in 1964

The Early Stages

Over 50 Years Ago

Part of Whiteshell's fleet of cars.

WR-1 Reactor Vessel

Installing the tunnels to connect the Whiteshell buildings

Bob Robertson

Tony Sawatsky

Bernie Pannell

Peter Dyne

Tom Tabe

Jack Remington

Keith Chambers

Max Allan

Alex Mayman

Ray Wazney

Ed Wuschke

William Ayres

Colin Lennox

Grant Unsworth

Roger Smith

Bob Pollock

Jim Putnam

Bob Lidstone

Harwant Singh

Cy Seymour

Warner Brown

Jim Harding

George Snider

Des McCormac

Memories of Whiteshell Laboratories
By Harwant Singh

In early 1965 Ajit Singh and I started to look for a place to work, after finishing our postdoctoral fellowships in Chicago at Argonne National Lab, and Northwestern Medical School, respectively. We applied to Radiation Labs in India, Canada, and the UK. We sent one set of applications to AECL Commercial Products in Kanata, in response to their advertisement in Science. They forwarded our applications to Whiteshell, and we were asked by Ray Kirkham in June to come for interview.

I was interviewed by John Weeks, and Abe Petkau. We were both impressed by the planning at the labs, and the planning of the town of Pinawa. This led to our reporting for work at Whiteshell, on Feb. 2nd, 1966, about 3 weeks before our first blizzard in Pinawa.

Abe Petkau's group was an energetic one to work with. I received good support from him, and from John Weeks. I was particularly happy to have Ara Mooradian's warm support, whose office was in our building at that time. My work included areas of interest for Medical Biophysics, starting with cellular components of protein synthetic machinery, and later in Radiation Applications. My earlier work led to collaborations with faculty members at universities in Manitoba and Oregon. Two memorable areas in later work were (i) oxy-radicals, and (ii) food irradiation. The International Conference the Medical Biophysics group (initiated by Ajit Singh) arranged in 1977 has been regarded by many as the basis of the emphasis on the use of anti-oxidants in foods, and as medicinal supplements. My work in Radiation Applications included providing technical support for applications for the approval of irradiation of various meats, to deal with bacterial contamination, in Canada.

Over the years it was nice to see increasing contributions at Whiteshell by women, including scientists, engineers, technologists, library personnel, technical editors, and the supporting staff (secretarial staff). It was historical, indeed, when one of them (Eva Rosinger) was made a senior manager. Some of the women who contributed to work at Whiteshell appear in the pictures, in this history.

Harwant Singh

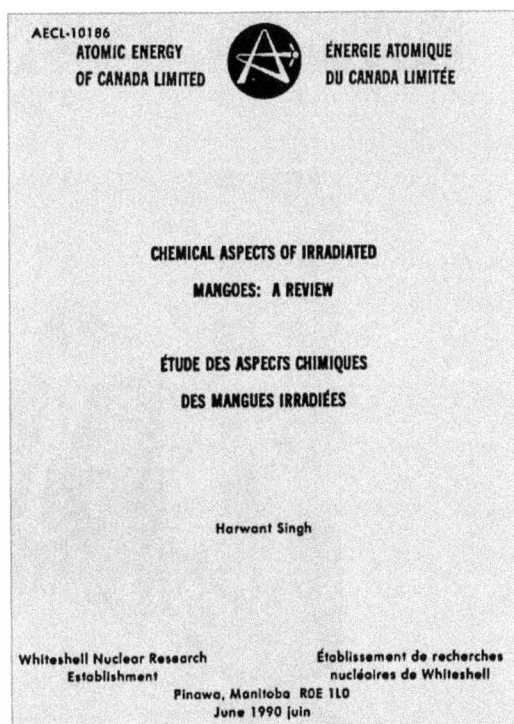

AECL-10186

ATOMIC ENERGY OF CANADA LIMITED

ÉNERGIE ATOMIQUE DU CANADA LIMITÉE

CHEMICAL ASPECTS OF IRRADIATED

MANGOES: A REVIEW

ÉTUDE DES ASPECTS CHIMIQUES

DES MANGUES IRRADIÉES

Harwant Singh

Whiteshell Nuclear Research Establishment

Établissement de recherches nucléaires de Whiteshell

Pinawa, Manitoba ROE 1L0
June 1990 juin

WNRE – mid 1960s

Bill Hancox, Mitch Ohta, Tom Tabe and John Hilborn reviewing facility
drawings

Front Row - Ara Mooridian, Len Simpson, Dave Faulkner, Roy Styles, Larry Simonson, Sid Jones, Ed Sexton, Bob Robertson; Middle Row – Ron Stevens, Ken Witzke, Roger Dutton, Ralph Mills, Ralph Moyer, Unknown, Tom Clausen, Lloyd Wiebe; Back Row - Abe Unger, Al Rogowski, Mike Wright, Al Reich, Brian Wilkins, Alex Wasylyshyn, Garnet Marks, Kurt Sprungman

WNRE Parking Lot – mid-1960s

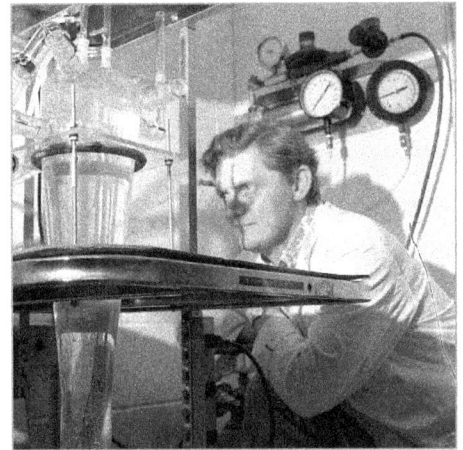

Lorne Swanson

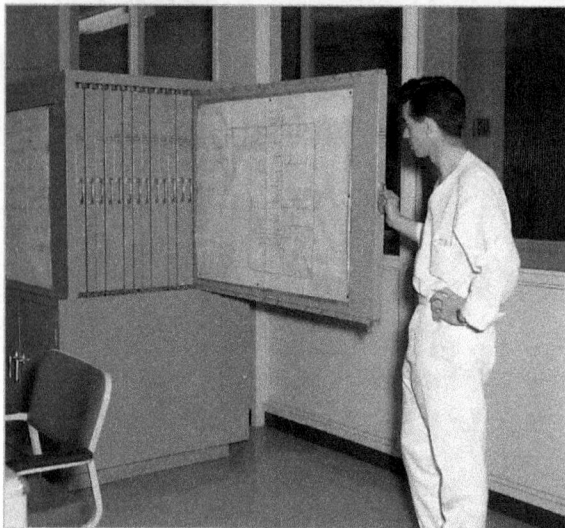

Thor Borgford checking valve arrangement on
flowsheet – 1965

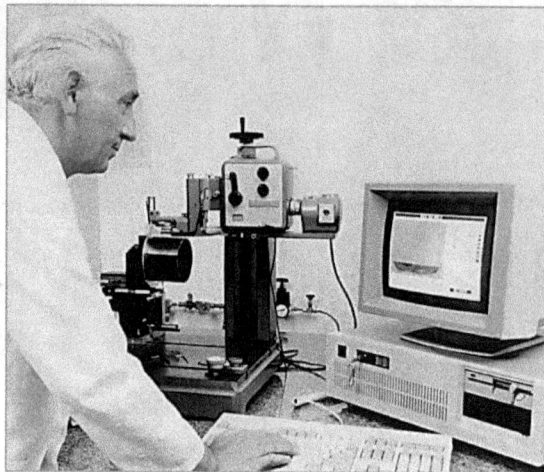

Frank Hughes testing samples

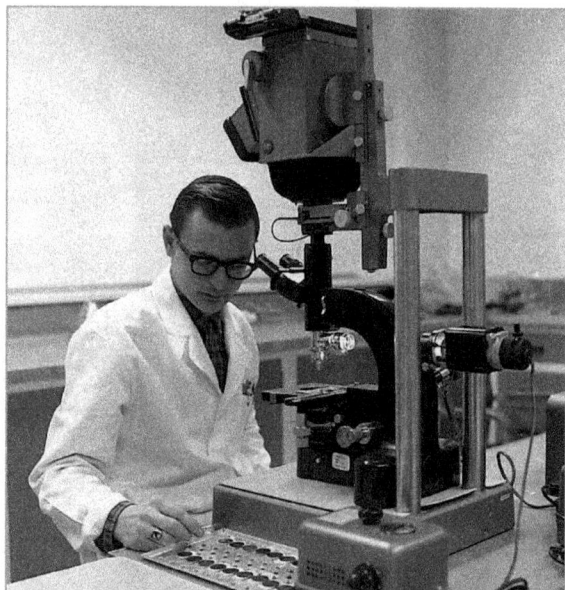

Garry Sherman using high-powered microscope –
1966

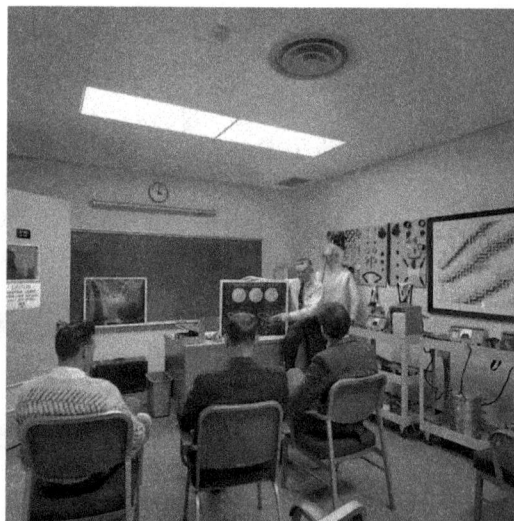

C.H. Plunkett explaining the structure of
hydrogen isotope

Roy Barnsdale and Jack Remington preparing
new WR-1 fuel strings for storage – 1965

Gerry Hampton

Building 300 machine shop

Gerry Smith

Whiteshell Memories - How Radiation Chemistry Began At WNRE
By Michael Tomlinson

The Right Honourable John G. Diefenbaker was the prime mover of Whiteshell Laboratories and the radiation chemistry work therein. He was from Prince Albert, Saskatchewan and when re-elected with a majority government in 1958, he decreed that the prairie regions should share in whatever benefits might be had from atomic energy. The outcome was Whiteshell Nuclear Research Establishment and the Local Government District of Pinawa (LGD). At the outset, the goal of WNRE was to build and operate an organic-liquid cooled nuclear research reactor to demonstrate that organic liquids could carry atomic heat from nuclear fuel to outside the reactor where it could be used to generate electricity at less cost than existing types of nuclear power plants. The task of the radiation chemistry group was to determine which organic liquid composition would work best and find out how the composition and properties would be changed by radiation and heat.

I first came in contact with AECL when I joined the ABC team (American, British and Canadian) as the UK chemist at CRNL in 1954. Dr. Bob Robertson supervised the team and treated me as part of his family. He was Best Man at my wedding to Helen Cox in 1959. In the 10 years from1949 to 1959, I had been finding out about the chemistry in most of the different kinds of nuclear reactors that were being developed in various countries in the world. These included the gas cooled reactors for the UK, (also France); water cooled reactors for the UK, US and Canada; and the European "Dragon Project" for a high temperature helium cooled reactor.

The first scientist to arrive at the nascent Whiteshell Nuclear Research Establishment was me. I arrived on August long weekend of 1963 with wife, Helen and first son, Martin, not yet age 2. WNRE and its town-site, Pinawa, were hectic with building amid churned up Manitoba gumbo. No homes were ready, so we were quartered at Casey's Inn in Lac du Bonnet, which was revelling at all hours with bush-workers and holiday makers. Bill McEwan, the housing manager did his special accelerator project and after

Michael Tomlinson

10 rumbustious days at Casey's we moved into row house #18 on McGregor Crescent in Pinawa, while construction hammered on in the rest of the row.

I signed in at the (not quite ready yet) WNRE. The only completed and operational building was the gatehouse and reception office. I was the early nerd, there to talk with the engineers, builders and trades-people about the completion and fitting out of the shielded room for the 1.5 million volt electron accelerator that was scheduled for delivery in the fall from High Voltage Engineering Corporation in Burlington, Massachusetts.

The shell of the main research laboratories (known as Building 300) was up and work on the interior structures had just begun, with the accelerator room as top priority. My lair was down in the basement near where many tons of concrete shielding were being poured. It was my singular privilege to launch into service the first toilet in the research building. The toilet provided minor mountaineering exercise because it was three flights up to the top floor and to the diagonally opposite corner of the building. It was labelled Ladies.

Bob Robertson became the first Director of Research for WNRE, and he gave me a 1.5 million volt (MeV) electron accelerator to play with. My job was to get it installed and use it to find out the chemistry that would happen inside the new WR 1 reactor that the engineers were starting to build. Accompanied by adept and diligent colleagues, we were able to use the accelerator to squirt energetic electron radiation into prospective reactor materials more than ten times as fast and intensely as the most powerful research reactors at Chalk River Nuclear Laboratories (CRNL) or elsewhere. This allowed us to leapfrog ahead to ensure that the new type of reactor, WR 1 would perform according to expectations and keep on doing so for the next 25 years. In particular we were able to figure out how to get optimum performance from the fresh and used organic coolant mix that was the key new feature of WR 1. By 1965 WR 1 was up and running and it continued to operate, at least as well as originally envisaged, for the next 25 years.

Much of the early history of radiation chemistry at Whiteshell is recorded for posterity and public access in the published papers of the early participants. The journal publications document accurately (much more so than ancient memory) the experimental work we did and what we learned from it. The research work extended from 1962 until 1965, by which time the WR 1 reactor was built and running. The publications lagged by a year or more behind the analysis and write-up of the work, so they extend from 1964 to 1968.

For any reader seeking more information, there is with me (or my estate) an archive of preprints or copies of all documents that I co-authored in the early WNRE period, 1962 to 1965 and beyond. Indeed, the archive covers all work that I co-authored from 1949 to 1985 including most of those from the UK in 1949 to 1959, which were initially secret for reasons of commercial and national security, i.e. not published. Before moving to AECL in 1962, I was able to arrange declassification of these UKAEA documents, which were then available from Her Majesty's Stationery Office at P.O. Box 569, London S.E.1. There are copies in my archive.

There were no secrets, except temporarily in the UK as noted above. We wrote up our findings and published them in science and engineering journals. However, the story was rather intricate and not newsworthy. So journalists, broadcasters and TV crews portrayed WNRE as a place of sinister secrets so as to be quite saleable by the fifth estate.

Gifted Colleagues and Amiable Acquaintances Who Made it all Happen

John Wright and Peter Davidge of the U.K. Atomic Energy Research Establishment (UKAEA) Harwell are the grandfathers of radiation chemistry at WNRE. John, my first boss, taught me all the basics of radiation chemistry, atomic power and much else when I took my first job there in 1949. Peter was the liaison man between John's section and people at other sites who were keenly interested in our findings.

Dr. A.W. (Bill) Boyd is the founding father of radiation chemistry at WNRE. When I joined AECL in November 1962 at CRNL, preparation of WNRE was just beginning. I joined the Chemistry Division to be taught by Doctor Boyd, who was already working on the radiation chemistry of prospective organic coolants. Thanks to Bill, I got a 9 month introduction to equipment and techniques for irradiation of organic coolant liquids in the NRX reactor. I learned how to measure the resulting changes in chemical composition and physical properties on his equipment, which I proceeded to duplicate.

Bill's very able assistant, Grant Bailey was my mentor and guide to lab apparatus and techniques. He kept the experiments going while I was often distracted by preparations for the Van de Graaff accelerator and planning research for next year. It was my blessing that Grant grew up in Portage La Prairie, MB, so he came to WNRE, the first to join the Radiation Chemistry group. He and his family have been a major presence in my life in Canada.

Ed Engel was the first Accelerator Operator at WNRE. He joined me in 1963 to learn how to assemble the newly arrived components of the accelerator and then operate and maintain it. Like many operators who followed, Ed came with a good training in electronics technology from the Alberta Institute of Technology. He was a newlywed and from a prairie farm background, so adapted well to Pinawa life.

Dick Poitier was an engineer who came from the High Voltage Engineering Corporation (HVEC) to direct the assembly, testing and running up to full power of the 1.5 MeV machine. When Dick arrived at WNRE in 1963 the only nearby accommodation was in the small motel with associated general store and restaurant at the Otter Falls Resort. The folks back at HQ in Boston lined up for his daily telephone report, not so much to hear about progress on the accelerator, but to hear Dick's tales of his encounters with the wildlife that came scrounging around the resort. The supreme highlight of his stay came when the resort's owners,

Steve Kondrachuk and his wife had to leave for some family event so they asked him to look after the Resort for a week. Dick's crowning moment was when he rented an equipment salesman a room, invited him to dinner at the restaurant and provided fill up with gas for his return trip to Winnipeg next day.

In 1965, Bill (W.G.) Unruh was the first summer student to contribute to early radiation chemistry at WNRE. Bill was completing his B.Sc. in physics at the U of M. He was a great asset. He computed energy absorption rates into selected elements when exposed to X-rays such as those generated with the 1.5 MeV electron accelerator. This work was published with M. Tomlinson as co-author in the journal "Nuclear Applications" in 1967. This publication was young Bill's first scientific publication with very many much more weighty ones to follow. At WNRE, he also contributed to computing of dose rates from fast neutrons and gamma-rays within the WR-1 reactor, which was just then being brought into service. Bill went on to achieve great academic distinction and has been for many years a very eminent professor of theoretical physics at UBC.

Dr. R.S. (Bob) Dixon was the first academically trained radiation chemist to join WNRE. Bob and his wife Rita arrived from England at Halloween 1966 just in time to join a very spirited throng of Pinawans in fancy dress. At this time, WR-1 was running routinely at full power and we were just beginning to tackle questions about the chemistry going on inside CANDU nuclear electric power stations. Bob's thesis work had prepared him to delve into the fundamental radiation chemistry of irradiated water. From this time onwards radiation chemistry branched out from organic reactor coolant studies in several directions. Dr. Dixon and other newcomers from academia applied the knowledge that they had acquired in their doctoral studies to expand basic radiation chemistry so as to further enlighten the quest for beneficial applications of atomic energy.

Accolades for WNRE Engineers and Artisans

The versatile and ever amiable engineer Jim Putnam was in charge of the accelerator room construction and much else. Names of the other participants are lost to me in a blur of new faces. However, I can and will pay tribute to all by describing the magnificent door that they provided for the accelerator room. Up to that time the standard entry-way to a shielded room for high radiation machines was a labyrinth of progressively thinner concrete walls with several turns that the radiation couldn't bounce around. At the outer end they had a final light steel grill, to control access. Jim's team convinced me that they could build a much more compact door that would save space, entry time, etc. They were eager to show that forward thinking and originality was already taking root at Whiteshell. The door was built in short order as a one-piece door of steel-reinforced high density concrete. The ten-ton door trundled open or shut, on steel wheels on rails, as easily as entering your favourite pub. It served well the many radiation chemists and other scientists who studied chemical and physical changes of irradiated materials there.

We are also greatly indebted to engineers, skilled artisans and glassblowers for devising and making containers and controllers for our irradiation experiments. Peter Kingston's machinists in the workshops of the R&D labs were wizards at turning our rough sketches into irradiation vessels and devices, even though they often had to be made with unusual and difficult materials.

To meet the greatest challenge of all, Ray Sochaski, P. Eng., designed and had built a loop on wheels, which could be trundled into the accelerator target area where it could simulate a reactor coolant circuit. An electromagnetic beam-modulator wiggled the electron beam up and down to spread the narrow, maximum power, beam of electrons into a target tube of the loop which had hot organic coolant flowing inside. The nickel alloy walls of the tube had to be very thin to let the 1.5 MeV electrons pass through. The loop trolley had upon it the pumps to circulate the organic liquid, and the heaters and controllers to maintain the temperatures that would be experienced in a reactor. For loop maintenance, or to free the accelerator experiment area for other users, the trolley was disconnected and wheeled out into the general laboratory area. The entire rig was very different from anything encountered in normal engineering practice.

The loop was essential for exploring certain aspects of irradiated coolant behaviour. Much of the exploratory radiation chemistry used steady irradiation at fixed dose rates. In the loop, as in the proposed reactor, the coolant was irradiated intermittently. Each molecule or slug of flowing liquid received a pulse of radiation as it passed through the reactor core and heated up on the way. Then it cooled and had no radiation when it passed through the pumping and control equipment outside the reactor. We had already surmised and confirmed that, for basic chemical reasons, the cooling off period allowed the coolant decomposition to proceed somewhat differently than for continuous irradiation. Ray's loop allowed us to take the measure of these pulse-irradiation effects on coolant decomposition and properties.

The Mutual Benefits of Reactor Coolants and Beer

Beer is an organic coolant that is popular with summer sports enthusiasts such as Dr. D.W. (Dave) Shoesmith, former WNRE electro-chemist, now Distinguished Professor at the University of Western Ontario. Early British nuclear power reactors were popular with the British working man because they kept down the cost of beer. The coolant for the Magnox and AGR (Advanced Graphite Reactor) types was carbon dioxide bought in large quantities from the brewing industry which generated huge quantities as by-product from the fermentation of barley to make beer.

Organic coolants for nuclear reactors are very different by-products of the citrus fruit growing industry. The chemical biphenyl is sprayed in the orange groves of Florida and citrus farms everywhere to protect the ripening fruit from attack by fungus and other diseases. The by-product terphenyl is produced and separated in major quantities in the purification of the biphenyl. Monsanto Chemical Co. promotes terphenyl-based liquids as organic coolants good for use in equipment that must operate at high temperatures. Thus organic reactor coolants encourage the consumption of that alcohol free, health promoting coolant - orange juice.

After Words

In 1969, when the 1.5 MeV machine was to be replaced with a 4 MeV Van de Graaff, I arranged for the 1.5 MeV machine to go to Professor Dave Armstrong at the University of Calgary to enhance his radiation chemistry capabilities there. Present at the inaugural ceremonies was Calgary Professor Van de Graaff, grandson of the inventor of the first such machine. Thus the ceremony was somewhat a homecoming for WNRE's revered first machine.

Our new 4 MeV electron accelerator incorporated recent advances in technology and capabilities. It had the latest in pulsed electron sources, a capability that I had lusted for ever since about 1960, when I attended a talk by an eminent Yorkshireman and chemistry graduate of the University of Leeds, George Porter. He told us about his research with flash-photolysis, which he had invented to enable him to observe chemical changes over very short time scales. Flash-photolysis revealed intermediate reaction products forming and then transforming into the stable end products. Such changes happened to the film inside your camera in 1960, in the blinking of your eye after the flash bulb went off. Flash-photolysis was a great step forward in learning about chemical (and physical) kinetics.

After this breakthrough, the quest began to apply the same intense pulse technique with radiation of higher energy than visible light. The shortest time scale I could achieve with the 1.5 MeV accelerator in the sixties era was mere milliseconds. In 1969 our new director of research, Dr. Peter Dyne, already a seasoned radiation chemist, obtained funding for a new 4 MeV Van de Graaff electron accelerator that was capable of pulse-radiolysis in the nanoseconds time scale. This enabled us to work at the forefront of research in radiation chemistry again. Moreover, Peter enabled the hiring of several more academically trained radiation chemists who joined Bob Dixon in expanding the scope of basic radiation chemistry that was relevant to the growing uses of atomic energy. During this phase we had several post-doctoral fellows share their knowledge and skills with us for two years at Whiteshell before moving on to stellar careers elsewhere. One of them was Mike West, who went on to the Royal Institution on Albermarle Street in London UK when

George Porter was President. Mike's new workplace was where Michael Faraday made many of his discoveries about electricity and magnetism in the 19th century.

At WNRE, Bob Dixon worked with the 4 MeV machine from 1968 onwards with this talented group of radiation chemists. Several years later the 4 MeV machine was equipped with the option of a proton source to enable study of radiation damage of materials by an intense proton beam. A 10 MeV irradiator incorporating new linear accelerator technology was introduced by AECL in 1987 for commercial mass-production of irradiated products. A pilot plant was set up at WNRE in a new building (B305) by the Radiation Applications Research Branch headed by Stu Iverson.

By 1990 the shut-down and decommissioning of WNRE was beginning. In 1998 the 10 MeV linac building and contents were sold off to an enterprising group of WNRE employees, who founded Acsion Industries. The performance of Acsion Industries over the past 17 years and their prominent showing in North America and beyond is truly awesome. I applaud Acsion's onward progress into the future as the most eminent atomic energy business within Pinawa LGD. It is a very heart-warming sequel to my installation of the first, 1.5 MeV irradiator and launch of the early days of radiation chemistry at WNRE in 1963.

Peter Cliche

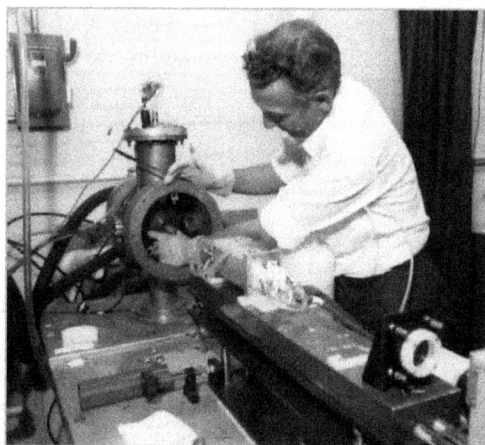

Frank Hughes

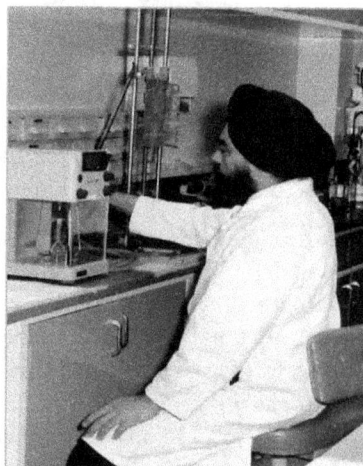

Ajit Singh

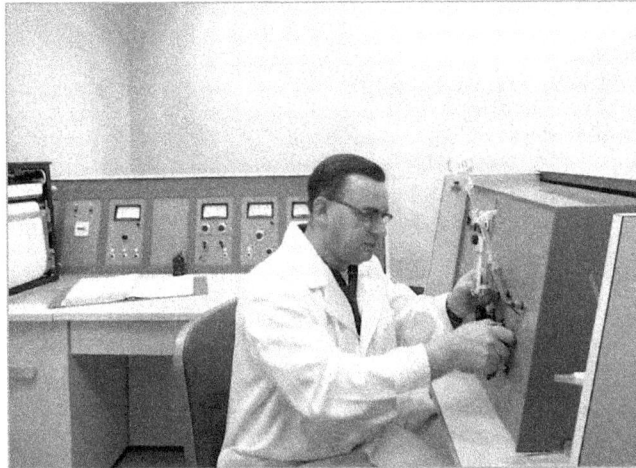

D. Wells in the mass spectrometer laboratory.

Bob Pollock in the WR-1 Control Room – 1965

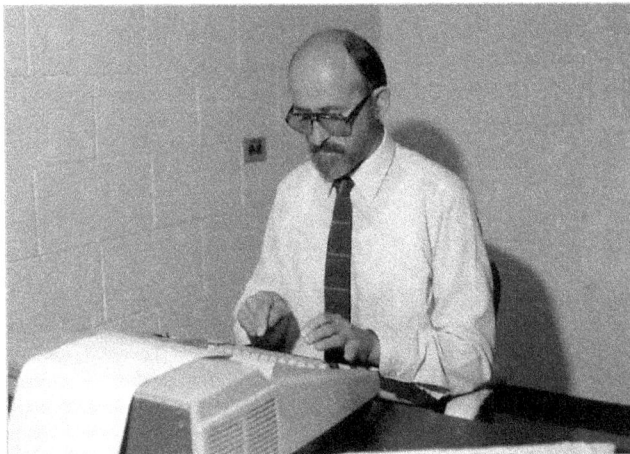

Reto Zach

Woman's Association at Whiteshell
By Chris Saunders and Anita Drabyk

Ensuring equal opportunities for women in science has been an issue for many decades. Even after some graduate schools opened their doors to women, not all were so welcoming. Indeed, Princeton, New York University, and Harvard did not grant women PhDs until the 1960s. As was true over and over again, as some barriers fell, others arose. For example, as the marriage ban was phased out (as was the pregnancy ban), anti-nepotism rules took hold. A woman scientist married to a male scientist would not be hired by the same department and was often refused employment at the same employer. The result: many highly educated married women scientists could not find employment.

As anti-nepotism rules came and went, the two-body problem arose and is still a barrier to women's advancement in science. What is the two-body problem? Simply stated, when one partner of a scientific or engineering couple is offered a position, the other spouse (typically the woman) also needs to find appropriate employment. Rarely is it possible to place two high-level academics at the same institution. More often than not, the female is offered no position or a position at a lower level than the one she left. Thus relocating is often beneficial to the male in the couple and detrimental to the career advancement of the "trailing spouse."

AECL struggled with the idea of women in scientific, engineering and senior management positions in the 1950s and 60s. In fact, W.B. Lewis was quoted as saying that he would not hire a woman for a scientific position (Nucleus – The History of Atomic Energy of Canada Limited, Page 110). Women occupied many of the support and administrative positions at Whiteshell from the mid-1960s. However, with a few notable exceptions, over the first 20 years at the Whiteshell site, the scientific, engineering and management staff were virtually all men. This began to change slowly in the 1970s and 1980s.

Excerpt from Robert Bothwell's *Nucleus – The History of Atomic Energy of Canada Limited*, Page 110, 1988:

"For twenty-five years, Lewis, through character and intellect, dominated Chalk River. When he was absent, he yearned to return, especially if a conference or a seminar was pending. By the late 1950s, his staff were known to rehearse what they had to say. If a new boy (for in Lewis' shop no women were scientists) had the jitters, he could run through his lines with his friends taking the part of Lewis from the floor. Success was measured in accurate anticipation of the director's comments and questions; failure was not to be thought of. The fact that Chalk River was an isolated society, remote even from science departments in universities, added to the psychological pressure. To many, the pressure meant excellence; to not a few others, though a minority, it meant intolerable egotism."

As more woman were hired, issues arose that needed to be addressed; new facilities such as washrooms, access to properly fitting clothing, lab coats, rubber boots, gloves, and safety equipment, and the need for managers and employees to be aware of and trained to handle issues around sexual intimidation and harassment. Managers were also expected to ensure all staff, including women, had equal opportunities for career advancement.

Progress was slow in many cases with woman employees at Whiteshell either tolerating a particular situation or leaving the company. In the early 1980s, the Women's Association was formed to be a unified voice to push for the needed changes to improve working conditions and to lobby to make sure all women at Whiteshell were given equal opportunities as their male co-workers.

Whiteshell has seen progress over the past 30 years. Today, the number of women in scientific, engineering and management positions is growing. Joan Miller, AECL's Vice President of Waste Management and Decommissioning, was the first woman responsible for the operation of the Whiteshell site. Whiteshell has a comprehensive process to train employees and managers and to address employee complaints. However, gender representation remains uneven, with men still outnumbering women at all scientific, engineering, and management levels. Even today, women scientists' chances for advancement are relatively poor. Women also, on average, earn less than their men counterparts.

The under-representation of women in the sciences and engineering has important consequences at all levels: for women, educational institutions, and businesses. Take the low numbers of women in decision making roles. To the extent to which managers and administrators oversee institutional (financial, human capital) as well as strategic resources, they also influence scientific knowledge and production. It follows that not only do women's perspectives and contributions in the sciences remain underappreciated but gender inequality at the very top can have significant consequences for the growth of scientific knowledge, productivity, and profitability.

Research on the dearth of women in STEM (science, technology, engineering and math) has mostly focused on the pipeline issue: namely, the more women you get into undergraduate classes, the more will come out the other end. National enrolment is only 19%; just 12% of the Canada's 280,000 professional engineers are women. Even if they do get a job, the number of women in leadership roles is just as low.

For example, at universities, only 12% of full professors in STEM are female, according to 2009 data from Statistics Canada; they are more likely to be working as contract faculty or as assistant professors.

When it comes to recognizing a body of work with an award or a prize, the numbers are just as discouraging. Only 11 of 60 members of the Canadian Science and Engineering Hall of Fame are women; 22 out of 186 prizes worth more than $200,000 were given to women by the Natural Sciences and Engineering Research Council (NSERC) between 2004-14; and 23 out of 202 people named to the Royal Society of Canada's Academy of Sciences in the past four years were female.

Eva Rosinger

Eva Rosinger was a long-time Pinawa resident as a scientist and Senior Manager with AECL. Eva's academic degrees are a M.Sc. in Chemical Engineering and a Ph.D. in Chemistry. She is the author of more than 40 scientific paper and articles, two patents and numerous conference presentations. Eva has broad interests in the arts, culture and sports, as well as social and health issues. From 1994 to 1998, she was Deputy Director for Environment at the Organization for Economic Co-operation and Development (OECD) in Paris, France. Dr. Rosinger was Director General and CEO of the Canadian Council of Ministers of the Environment (CCME). Dr. Rosinger is a member of the Public Advisory Panel for the Environmental Commitment and Responsibility Program, Canadian Electricity Association and of the Lectures Committee of the Royal Canadian Geographical Society.

She is a past President of the Canadian Nuclear Society and a former member of the Board on Radioactive Waste of the US National Academy of Sciences, the Board of Directors of the Canadian Institute of Child Health, the Board of Directors of the Winnipeg Symphony Orchestra, the Council of the Association of Professional Engineers of Manitoba and a former vice-chair of the Advisory Development Board for the Banff, Yoho and Kootenay National Parks. Dr. Rosinger is a recipient of the 1992 YM-YWCA Woman of Distinction Award, the 1988 Certificate of Merit by the Government of Canada for Contribution to the Community, and the Order of Sport Excellence and Achievement Award by the Government of Manitoba. She is listed in Canadian Who's Who, Who's Who in Canadian Business, American Men and Women of Science, and International Who's Who of Professional and Business Women.

A recent U.S. survey of female scientists in the workplace confirmed the existence of a career pipeline that steadily leaks qualified female candidates due to inherent and persistent biases. Women had to provide more evidence of their capabilities than male colleagues in order to receive the same recognition; 64% had their commitment to work questioned and opportunities dry up after they had a baby; and 35% reported being sexually harassed at work at least once. The women who do tough it out are consistently passed over for recognition.

Joan Miller

For 30 years, Joan worked for AECL and CNL at the Chalk River Laboratories. As the Vice-President and General Manager of Decommissioning and Waste Management, her main job was to lead people in the execution of a key component of the Chalk River and Whiteshell sites' activities.

Joan completed her B.Sc. (Hon.) in Chemistry from McMaster University in 1979 and immediately started with AECL. Throughout her career she was under the direction of many great supervisors who went out of their way to create new learning and development opportunities for her. They allowed her to take on significant activities and roles early on in her career, giving her the opportunity to interact with a large number of people within and outside the organization. They also provided a broad range of projects to work on, along with many activities from within those projects such as technical activities, regulatory interface, contract negotiation and customer presentations.

As a vice-president, Joan loved the problem solving aspect of her position. She also had the opportunity to influence program direction and took great satisfaction in watching her staff develop and accomplish their goals. Joan's most rewarding project was turning her Research and Development experience into a commercial facility that met a customer's need, along with the transfer of that R&D knowledge to the customer's engineering and operations staff.

There are some signs of institutional changes. The Canada Research Chair program published new guidelines for reference letters that provided tips on how to ensure unconscious biases don't undermine female candidates. Referees are encouraged to keep feedback specific to the position, and avoid adjectives that characterize women as maternal or agreeable. But that doesn't change the fact that women are paid less to do the same work, asked to do clerical work even if they are qualified professionally, seen as disagreeable if they speak up at meetings, and talked over when they do speak up.

The Women's Association at Whiteshell became part of the broader Whiteshell Association in the early 2000s. They continue to advance the issues important to women at AECL and CNL. They are assisted in this effort by a number of groups.

WiN (Women in Nuclear; www.wincanada.org) is a world-wide association of women working professionally in various fields of nuclear energy and radiation applications. WiN-Canada was formed

Researcher Donna Wuschke

in early 2004 and has been working to emphasize and support the role that women can and do have in addressing the general public's concerns about nuclear energy and the application of radiation and nuclear technology. WiN-Canada also works to provide an opportunity for women to succeed in the industry through initiatives such as mentoring, networking, and personal development opportunities.

Since 1981, the Society for Canadian Women In Science and Technology (SCWIST; www.scwist.ca) has focused on encouraging women into science, engineering and technology. Outdated assumptions persist about women as leaders in science, engineering and technology. Men continue to dominate senior leadership positions within these areas, despite the equal ability of their female colleagues. SCWIST supports and promotes the education of girls and women through programs and activities that they develop in partnership with the community. They boost the numbers, retention and status of women in the workplace by facilitating networking, mentoring and advocating woman-friendly policies. They highlight opportunities, achievements and positive messages for and about women in the field. We do this by raising public awareness and guiding policy implementation.

The Canadian Centre for Women in Science, Engineering, Trades and Technology (WinSETT Centre; www.winsett.ca) aspires to recruit, retain and advance women in science, engineering, trades and technology (SETT). Through collaboration and partnership, the WinSETT Centre creates and fosters opportunities that encourage women to enter, stay and grow in SETT careers with the goals of maximizing Canada's human resource potential, increasing innovation, and driving Canadian economic development.

Promoting careers for women in the natural sciences and engineering is also a priority for Natural Sciences and Engineering Research Council of Canada (NSERC; www.nserc-crsng.gc.ca). They are committed to increasing the number of women in these fields, facilitating the accommodation of career and family, and nurturing mentorship. The Chairs for Women in Science and Engineering Program (CWSE) was launched in 1996. Its goal is to increase the participation of women in science and engineering, and to provide role models for women active in, and considering, careers in these fields. The program is regionally based, with one Chair for each of the Atlantic, Quebec, Ontario, Prairie, and British Columbia/Yukon regions.

As a national organization of groups, institutions and industries, the Canadian Coalition of Women in Engineering, Science, Trades and Technology (CCWESTT; www.ccwestt.org) promotes girls and women studying and working in these fields. It celebrates the contributions of women in all spheres of Engineering, Science, Trades and Technology, from education to retention and leadership. CCWESTT has over 20 member organizations, each focused on issues related to women in scientific, engineering and management careers.

Lorna Lubitz at the library

Janet Dugle

Janet Dugle was born in Pierre, SD. She grew up and attended high school in Dupree, SD, then attended Carlton College MN, Yale University CT, and the University of Alberta, Edmonton, where she received her Ph.D. in Botany.

Jan worked at AECL in the Environmental Research Branch starting in 1967. She had a large research project in which she studied the effects of gamma irradiation on plants.

Jan was an avid photographer and was involved with the Pinawa Photography Club. She was also a member of the Trans Canada Trail committee providing her botanical and ecological expertise. She was instrumental in making the Ironwood Trail. She was also a member of the first Pinawa Art 211 group. Jan and Bill Murray started a framing and art business in town. Many a retirement gift or farewell gift was framed here.

Jan was part of the Pinawa Support Group who met on a weekly basis. There she found friends and travelling partners who were an important part of her life. An avid bridge player, teacher, and director at the Pinawa Club, Jan achieved her bronze life masters. She enjoyed meeting people and travelling to tournaments. She was pioneer survivor with an artificial heart valve for 33 years and then had another replacement in 2012.

Women at Whiteshell – 1968; 89 Employees - 13% of Staff

Allwright, Mrs. E.C.
Armstrong, Mrs. L.A.

Baird, Miss L.C.
Belinski, Mrs. P.M.
Bell, Miss E.J.
Benson, Mrs. S.M.
Benton, Mrs. A.B.
Buss, Miss L.V.
Butchart, Mrs. L.
Butler, Mrs. E.

Carlson, Mrs. D.D.C.
Christoff, Mrs. E.V.G.
Chura, Mrs. S.L.E.
Clarke, Mrs. G.M.
Clarke, Mrs. M.A.
Clark, Mrs. V.L.
Curie, Mrs. P.D.

Dhont, Miss J.E.

Dreger, Mrs. L.A.
Dugle, Dr. J.R. (Mrs.)
Dunn, Miss L.M.

Eisenzimmer, Miss M.

Fay, Mrs. S.D.
Fisher, Miss T.
Fitzsimmons, Mrs. M.E.
Frank, Mrs. S.A.

Hall, Mrs. B.J.
Hammond, Miss K.J.
Hanley, Mrs. S.R.
Harper, Miss L.G.
Hawley, Mrs. N.J.
Hopko, Miss B.
Hunt, Miss M.

Johnson, Miss, L.M.

Kabaluk, Mrs. C.A.
Karklin, Mrs. D.

King, Mrs. S.
Klepatz, Miss I.D.
Klimack, Mrs. J.L.
Kowalchuk, Mrs.

Lubitz, Miss L.C.

MacFarlane, Mrs. J.C.
McCarthy, Miss M.E.
McGinnis, Miss S.A
Magura, Mrs. J.M.
Marek, Mrs. D.C.
Mathews, Mrs. R.J.
Matsumoto, Mrs. C.
Mehta, Mrs. R.
Minton, Miss, E.J.
Mitosinka, Miss V.A.
Montague, Mrs. R.M.
Mowat, Miss B.B.
Murray, Mrs. J.P.

Oleskiw, Mrs. K.

Ostryzniuk, Miss D.C.

Perritt, Mrs. M.M.
Pleskach, Mrs. J.C.M.
Plotnikoff, Mrs. R.M.
Pouteau, Miss R.M.L.
Presser, Mrs. L.D.

Ramsay, Mrs. A.M.
Reavley, Miss J.

Saarela, Mrs. P.M.
Schinkel, Mrs. E.
Schick, Mrs. B.J.S.
Schmidt, Mrs. C.
Schwartz, Mrs. O.C.
Scott, Mrs. V.A.
Shierman, Mrs. M.C.
Singh, Dr. H. (Mrs.)
Sobetski, Mrs. B.C.
Solberg, Mrs. W.C.
Stadnyk, Mrs. A.M.

Steffes, Miss L.M.

Taylor, Mrs. O.M.
Theriault, Mrs. R.
Ticknor, Mrs. J.S.
Tirschmann, Mrs. J.D.
Trask, Miss R.

Urbanietz, Mrs. J.E.
Urbanski, Miss G.

Waslyshyn, Mrs. S.
Wasywich, Mrs. V.L.
Weissig, Miss M.R.
Wood, Mrs. M.H.
Wuschke, Mrs. D.M.

Yakymin, Miss O.V.

Zieski, Mrs. G.L.

Women at Whiteshell – 1994: 274 Employees - 34% of Staff

Abraham, Cathy
Aikin, Charla
Aitken, Charla
Alexander, Gaille
Altstadt, Linda
Archambault, Tara
Arneson, Colombe
Arsenault, Karen
Arthur, Debra
Augustine, Carol-Lee
Ayers, Heather

Backer, Susan
Bailey, Bonnie
Ball, Joanne
Balness, Betty
Balness, Netta
Barnsdale, Adonna
Beauchamp, Cathy
Behnke, Edith
Benson, Shirley
Bishop, Candy
Blom, Diane
Boczak, Diane
Boivin, Johanne
Bonekamp, Sandra
Borgford, Barb
Boyle, Shirley
Bratty, Marilyn
Brennan, Yvonne
Brincheski, Margaret
Broadfoot, Myrna
Brown, Debbie
Brown, Linda
Brown, Merle
Bruneau, Crystal
Burgoyne, Donna
Burns, Rose
Bush, Angel

Cafferty, Lesa
Campbell, Sandy
Carlson, Leslie
Champagne, Kimberley
Chung, Minda
Clarke, Alicia
Colotelo, Catherine
Cote, Sylvie
Cribbs, Mary Ann
Crognali, Rosa
Crosthwaite, Christine
Cutting, Tanya

Dalzell, Shannon
Daymond, Louise
Desrochers-Alpers, Lise
Doern, Diane
Donnelly, Wanda

Dooley, Fay
Dormuth, Alice
Drabyk, Anita
Drake, Myrna
Dreger, Stella
Drew, Patty
Dykstra, Sharon

Edwards, Carole
Elcock, Judith
Evans, Lorna
Ewing, Lyn

Farrell, Wanda
Fiebelkorn, Agnes
Findlay, Carol
Flett, Debbie
Fraser, Rhea
Frechette, Agnes

Gauthier, Annette
Gillespie, Phyllis
Gmiterek, Teresa
Goodwin, Marlene
Graham, Lori
Gray, Barbara
Greber, Mary
Griffault, Lise
Guttman, Grace

Haacke, Kathleen
Halley, Julie
Hamon, Connie
Hampshire, Joan
Hampshire, Sheri
Hanna, Kathryn
Hansen, Ethel
Harding Shirley
Harris, Debbie
Harrison, Ruth
Harvey, Kay
Hatland, April
Hawkins, Janice
Heckert, Donna
Henschell, Eileen
Hiebert, Irene
Hiltz, Gisele
Hladki, Norma
Hnatiw, Joan
Hocking, Margaret
Hoffman, Janice
Hood, Darlene

Johnson, Ellen
Johnston, Joanne
Johnston, Laurel

Kaatz, Susan
Kearns, Judy

Keith, Darlene
Kellendonk, Cindy
Kendel, Lynn
King, Sandra
King, Sharon
Kingsland, Laurie
Kirk, Carolyn
Klepatz, Linda
Knox, Margaret
Kossman, Maureen
Kovacs, Karen
Kovari, Marilyn
Kroeger, Aggie
Kufflick, Dori
Kukurudz, Joanne
Kuzyk, Sharon

Laurin, Suzanne
Laverock, Martha
Lavoie, Connie
Legiehn, Olga
Lidfors, Alanna
Lim, Monica
Lincoln, Raquel
Litke, Cynthia
Lodge, Cindy
Lofstrom, Rosalie
Lortie, Bonnie
Lucht, Lisa

MacDonald, Maureen
Marciniak, Mary
Marohn, Susan
Martin, Heather
Mathews, Sandy
Matthews, Virgina
Mauthe, Karen
May, Dawn
McArthur, Loreen
McAuley, Gloria
McCalder, Terry
McConnell, Jodi
McCooeye, Pat
McCoy, Lydia
McDowall, Barb
McDowall, Bonnie
McDowall, Marg
McFarlane, Joanna
McIlwain, Heidi
Meade, Alice
Meads, Cheryl
Mellors, Joyce
Meyer, Siegrun
Meyers, Leslie
Miller, Maureen
Miller, Susan
Moir, Deborah
Moltyaner, Freda

Monaster, Roberta
Morrish, Donna
Murphy, Pearl
Murray, Joanne

Nguyen, Elisabeth
Nuernberger, Carol
Nuttall, Christine

Olchowy, Helen
Olchowy, Leslie
Oliveira, Gertie
Onofrei, Maria
Orvis, Carol
Ostick, Ingrid

Palson, Valerie
Pargeter, Patricia
Patterson, Lynn
Payne, Barb
Pedersen, Suzanne
Plett, Debra
Porth, Pat
Portman, Ann
Puls, Gloria

Ramsay, Pat
Ramsay, Sandra
Randell, Doreen
Reschke, Lorna
Rigby, Marilyn
Rochon, Ela
Rodych, Lori
Rohrig, Cheryl
Roper, Tabatha
Rosentreter, Jean
Ross, Deborah
Ross, Karen
Rosset, Brenda
Ruta, Marilyn
Ryz, Mary

Sabanski, Barbara
Salmon, Carol-Lee
Sanderson, Tracy
Sanipelli, Barbara
Sargent, Jane
Saxler, Wendy
Schinkel, Elizabeth
Schultz, Debbie
Schultz, Francis
Schultz, Lorraine
Seifried, Sherry
Sheppard, Marsha
Shewfelt, Betty
Simpson, Judith
Singh, Harwant
Smith, Anne
Snider, Jacqueline
Sobetski, Bev

Solnes, Pat
Solomon, Larissa
Soonawala, Angie
Spinney, Christine
Spitz, Mary
St. Denis, Bonnie
Staerk, Shirley
Stam, Jan
Stephens, Barbara
Stevens, Carman
Stokes, Margaret
Strandlund, Leslie
Strobel, Karen
Stroes Gascoyne,
Simcha
Styles, Anne
Sullivan, Pat
Swanson, Janice
Szekely, Carol

Tabe, Joyce
Tamm, Judy
Tarr, Donna
Tateishi, Miyoko
Taylor, Ann
Taylor, Sharon
Thompson, Carol
Thomson, Maryann
Ticknor, Joan
Turner, Dorothy

Veilleux, Julie
Veroneau-Jansson, Terri

Wallace, Sue
Walters, Agnes
Walton, Carol
Watson, Angela
Wedgewood, Joanne
Welz, Nadine
Wilgosh, Brenda
Wilken, Dorothy
Wilkins, Sylvia
Wilson, Gisele
Winchester, Joan
Wojciechowski, Laverne
Wold, Kathy
Wood, Lucille
Woodbeck, Kathy
Worona, Heather
Worona, Shannon
Wren, Clara
Wright, Esther
Wuschke, Donna

Young, Patricia

Zach, Marg
Zahorodny, Janice
Zirk, Denyse

Women Staff at Whiteshell – 2013: 89 Employees - 25% of Staff

Adams, Leah
Aitkenhead, Christel
Akpan, Uduak
Avanthay, Kim

Bachman, Lori
Barnett, Gina
Barnsdale, Adonna
Blowers, Christine
Bodley, Sue
Bruneau, Toni
Bush, Kelly

Chshyolkova, Tatyana
Conroy, Karen
Crognali, Rosa
Cummer, Jody
Cure, Nicole

Davis, Charene
Dooley, Fay
Drabyk, Anita

Edwards, Carole
Elkin, Cindy
Emond, Lisa
Erickson, Trish

Forbes, Kim

Gagnon, Noeme
Gervais, Joanne
Gmiterek, Teresa
Graham, Lori
Grzegorzewski, Julia

Hessian, Cindy
Hirst, Sheri

Jansson, Brenda

Keith, Darlene
Kryschuk, Chantelle
Kuhl, Sherri

La Rue, Nicole
Laurin, Suzanne
Lavallee, Wendy
Laverock, Martha
Leake, Lorraine
Leishman, Connie
Lindsay, Fiona
Luke, Adriana

Matheson, Sandi
McDonald, Fiona
McLean, Suzanne
Meeker, Pamela
Milhausen, Dawn
Mitchell, Melanie
Mlodzinski, Anna
Molinski, Tammy
Mynhardt, Sunette

Pachkowsky, Lori
Pargeter, Pat
Parisian, Sherry
Pawluk, Patricia

Radyastuti, Amina
Rasheed, Sana
Redmond, Angela
Reimer, Eleanor
Renard, Edna
Ritchot, Richelle
Rosentreter, Michelle
Rowlands, Jenn
Ruta, Jodi

Sieg, Joanne
Singbeil, Celeste
Smyrski, Sherri
Stefaniuk, Janet
Stefaniuk, Shannon
Stelko, Melissa
Stephenson, Kate
Struss, Ashlee
Swaenepoel, Shannon
Swain, Laurissa

Tapia, Elen
Taylor, Angela
Taylor, Louise
Thiessen, Kati
Tiede, Myrna

Wiebe, Carly
Wilcox, Alanna
Wilgosh, Brenda
Wilgosh, Monica
Wilson, Leslie
Wojciechowski, Laverne
Worona, Shannon

Xu, Daisy

Yule, Pamel

Sandy King, the first woman to manage Whiteshell's Apprenticeship Program, with Nick Ostach and Ken Urbanski

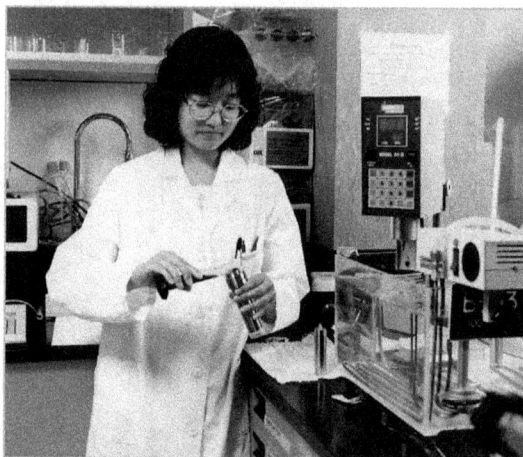

Researcher Miyoko Tateishi

Whiteshell Protective Services, Security and Fire Department – 1984: Front Row: Wes Jansson, Bob Early, Gilbert Smith, Fire Chief Tom Lamb, John Kowaluk, Bill McArthur, John Stefaniuk; Back Row: Norman Wold, Wayne McLeod, Bob Otto, Lawrence Germain, Ted Gmitrowski, Bob Thompson, Larry Carefoot, Harry Noel, Dave Beeching Jr., Jack Hildebrand, Kevin Maguire, Eugene Mamrocha, Dave Kearny, Hank Theunissen, Don Trudeau

Bruce Stewart

A retirement celebration for Ed Hawley – Glenn McGee, Barry Hood, John Findlay, Klaus Spitz, Thor Borgford, Robin Henschell

Keith Nuttall

Peter Sargent

Bill Hancox

Maple Reactor Research Team – Les Hembroff,
Dave Richards, Ken McCallum, Robin Henschell,
Janusz Kowalski

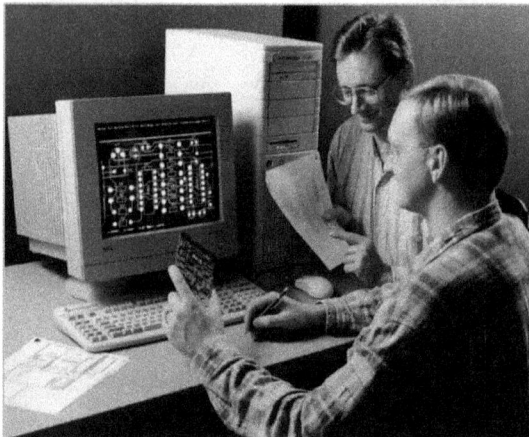

Willie Dueck and Brian Wilcox

Purchasing Staff - Darlene Beeskau, Carol Lee Augustine, Unknown, Maureen Miller, Al Peterson, Shirley Benson, Wendy Plishcke, and Pat Sullivan

Personnel Staff - Bud Mager, Maryann Thomson, Sandy King, Cheryl Rohrig, and Gord Magura

Technical Information Services Branch: Front Row - Aggie Walters, Mike Luke, Martha Tuxworth; Back Row - . David Haworth, Ray Karl, Siegrun Meyer, Ted Iwanowski, J. C. Leblanc

Public Affairs Branch: Back Row - Dave Studham, Angie Soonawala, Larry Christie, Bob Dixon, Metro Dmytriw, Marilyn Lloyd; Front Row - Christine Nuttall, Leni Dixon, Pat Blais, Fiona Wright

Service Awards at Whiteshell –Kurt Sprungman, Ray Rondeau, Thor Borgford, Tom MacDonald, Brian Wilkins, Leon Clegg, Ralph Moyer, Ken McCallum, George Penner, Les Hembroff, Frank Hughes, Bun Baxter

Andrzej Kosciuk, Sol Sawchuk, Dieter Jung, Gary Young and Greg Gowryluk.

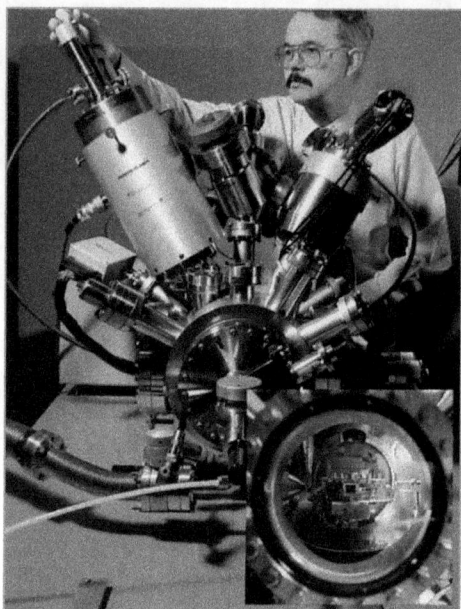

Bill Hocking using scanning auger
microscope combined with a secondary
mass spectrometer

Derek Owen using Fourier transform
infrared spectrometer

Terry Andres and Annette Skeet using
SYVAC

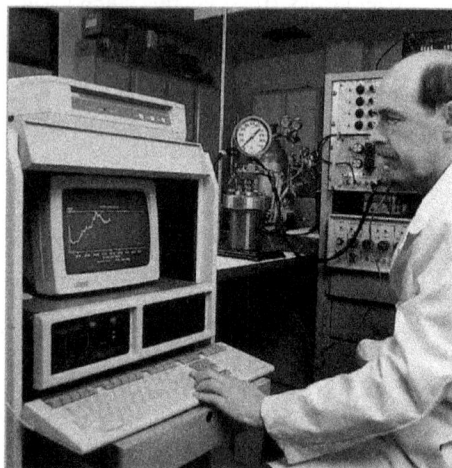

Grant Bailey studying corrosion
processes

1988 Whiteshell Safety Awards

Maintenance Branch – G. Sachvie, K. Jackson, L. Veroneau, S. Kekish, W. Sitar, W. Klapprat, L. Kroeker, A. Zerbin (presenter)

Radiation and Industrial Safety Branch-
D. Hladki, R. Lambert (presenter)

G. Buchanan, W. Wasylenko, L. Carlson, F. Carlson, F. Gryseels, F. Jabush, H. Steinleitner, D. King, D. Kubish, M. Wright (presenter), Horst Sommerfeld

Radiation and Industrial Safety Branch –
S. Niszchuk, R. Lambert (presenter)

Materials Handling Branch –
J. Rogocki, M. Grant (presenter)

Manufacturing, Maintenance and Material Handing Division – M. Wayne, E. Muzychka, A. Gesell, D. Kabaluk, R. Agland, R. Schultz, J. Garbolinski, E.D. Lidfors (presenter)

24 Year Award – R. Lussier, M. Wright (presenter)

Manufacturing Branch –G. Teirpstra, D. Platford, G. Krawchuk, R. Radons, P. Clarke, E. Hemminger, K. Meek (presenter), W. Hawryshko, G. Graham

G. Murray, K. King, M. Wright (presenter)

15 Year Awards – Maintenance Branch – J. Kostuik, L. Voss, E.D. Lidfors (presenter)

20 Year Awards – P. Grzegorzewski, S. Herzog, C. Bruce, J. Hannon, G. Wensel,
M. Wright (presenter), W. Dereski

Safe Drivers – F. Carlson, L. Voss, D. Kubish, A. Gesell, C. Bruce, E.D. Lidfors
(presenter), G. Hampton, E.L. Veroneau

Maintenance Branch –N. Bruneau, K. Edwards, R. Suski,
A. Zerbin (presenter)

YOUR PHOTOBADGE

WHITESHELL NUCLEAR RESEARCH

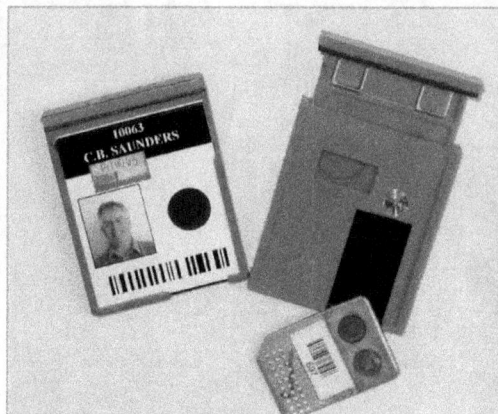

Do you remember your employee number?

Frank Walton

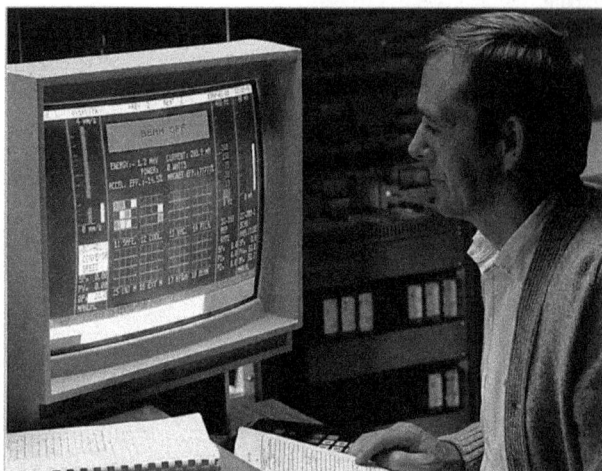

Wayne Stanley operating the I-10/1 linear accelerator

Alf Wikjord

Roger Portman

Dave Wren

F.W. Gilbert

A. Aikin

A. Mooradian

R. Hart

S. Hatcher

Protective Services – 1967: Front Row: Brian Monkman, Jim Arkle, Les Johnson, Maryann Thomson, Shirley Trask, Ben Banham, Tom Lamb, Dorothy Karklin, Walt Schmidt, Bob Otto, Bill Schwartz, Charlie James; Back Row: Bill McArthur, Neil Campbell, Neil Brown, Wayne McLeod, Gordon Kabaluk, Doug Taylor, Keith Lavine, Bob Early, Glen Thompson, Doug Taylor, Unknown, Norm Wold, Ted Gmitroski, Stan Penner, Vic Dreger

Reactor Safety Research Division - Front Row: Herb Rosinger, Rod MacDonald, Len Simpson, Lydia McCoy; Back Row: Bruce McDonald, George Gillespie, Keith Mayoh, Mani Mathew, Harold Wagner, Glen Lowry and Dave Richards

Reflections of a Career at WNRE and a Life in Pinawa
By Marvin Ryz

I can still remember reading the classified section of the Winnipeg Tribune one evening in the summer of 1963. I was looking for career opportunities, not that I didn't enjoy my work at the University of Manitoba Animal Science Department, where I assisted in various research projects, it's because I was looking for a career position, something that my job at the U of M didn't really offer. That's when I noticed the job application for research technicians to work at Atomic Energy of Canada Limited's (AECL's) newly created Whiteshell Nuclear Research Establishment (WNRE) as it was called at that time. The position was to conduct post-irradiation examination (PIE) of nuclear research reactor fuel and components. This work required that all operations be conducted remotely behind shielded rooms called Hot Cells. The Hot Cells were heavily shielded mini laboratories protected by up to 42" of ilmenite concrete and had viewing windows also up to 42" thick. I immediately thought this sounded pretty cool even though I had very little understanding of nuclear reactors or what PIE work was all about, let alone what a Hot Cell really was. It just sounded so futuristic; I had to submit my application for one of the positions being advertised. Much to my surprise and pleasure, a few weeks later I received a call to come to WNRE for a formal job interview. My first impression of the WNRE site was utter amazement at the furious pace of the work being conducted on site. Construction activity was taking place all over at this relatively little known facility along the Winnipeg River. There was a large construction camp site and camp cookhouse just to the south of the current site buildings. Impressive also was the amount of mud as this had been a very wet fall and given all the construction activity everywhere one went there was mud and more mud! This was also before the days of asphalt on the roads to either WNRE or the town site of Pinawa making these roads very rough and muddy as well. The WR-1 Research Reactor was still just a very large and deep hole in the ground. Only a few permanent buildings had been completed, one was building 402 where my job interview was held. The job interview just further convinced me that this was the opportunity I was seeking....research and intrigue. The job application must have gone well because a short time after the interview I was contacted to say I had been selected from over 200 applicants to be one of the Hot Cell technicians. The reason I believe I was selected was that one of the research projects at the U of M involved the use of the tracer radioactive isotope Iodine (I^{131}). It was put into the poultry feed and the migration of the I^{131} to the various vital organs of the poultry was monitored using radiation detection equipment. One never knows what key factor may lead to the next career move.

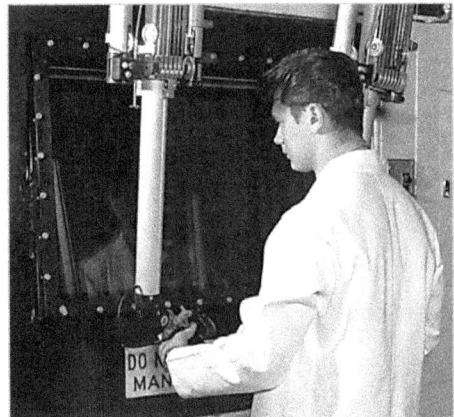

Marvin Ryz performing post-irradiation examinations in the hot cells in 1965

A condition of employment was that I had to travel to Chalk River Ontario, home of Canada's parent nuclear research site, the Chalk River Nuclear Laboratories (CRNL), where I was to undergo an extensive training period. This was an important period in my early career, as I learned the skills required to conduct the many remote PIE tasks that would be required at the WNRE Hot Cells upon my return to Manitoba. In 1965, after just over a year at CRNL, I returned to WNRE in time to assist in the commissioning of the first Hot Cells. This was followed shortly after by the achievement of first criticality in WR-1 in 1965 and the receipt of the first irradiated fuel assembly from WR-1 for PIE. I recall this historic day well as it attracted the attention of practically every employee on site. Most had never had the opportunity to see irradiated reactor fuel before, or watched how PIE procedures were conducted remotely in hot cell facilities.

I was one of a large number of young single employees at WNRE at that time. Since single employees were normally not eligible for either an AECL apartment or house, most were assigned a dormitory-like room in Kelsey House (now called the Wilderness Edge). At that time all the accommodations were owned by AECL. If one wanted to be assigned an apartment, row house or detached house they had to submit a request to AECL and assignment of accommodations was done based on a scoring system (based on need and qualifications). Accommodations, in the early days, were on a rental basis only, but in later years AECL made housing purchase available to employees. Many personal relationships developed either at WNRE or Kelsey House and this is where my wife Mary and I first met. Many of our friends also met in Pinawa and formed lifelong relationships. We decided to live in Pinawa and raise our children here. Pinawa provided the opportunity to enjoy a wonderful career, while living in a beautiful wilderness setting along the Winnipeg River. Wisely, the Winnipeg River shoreline along the town site was set aside as a public reserve area for all to enjoy. Pinawa provided opportunities for people with almost any pastime. Activities that did not exist naturally were quickly established by the many clubs and organizations that formed; one of the major ones that springs to mind and that exists today is the Pinawa Players. They have been entertaining locals for many decades with their first class performances. For me it was activities such as fastball, curling, snowmobiling, cross-country skiing, boating, and fishing. There were unlimited opportunities in these areas. The residents of WNRE and Pinawa also enjoyed many social activities. We can all fondly remember the Christmas parties in the 1960s and 1970s. Employees from several of the buildings at WNRE, such as those in Buildings 100, 300, 408 and 412, organized their own special Christmas parties. Other events much looked forward to were the annual Curling Club Awards banquets and the not to be missed classic Pinawa Birthday dances held at the Kelsey House.

Living in Pinawa also provided the opportunity for me to experience public service, first as a councillor for three years followed by nine years as Mayor of Pinawa, another experience that I found to be very enlightening and personally rewarding. Of particular note was the partnership between Pinawa and nearby community of Beausejour, to host the 1988 Manitoba Summer Games. Time has also seen the evolution of second and even third generation Pinawa residents making Pinawa and the Whiteshell Laboratories their home and career. Second generation residents have demonstrated an entrepreneurial spirit in developing trailer parks, housing sub-divisions and various business opportunities. Probably the only negative aspect of life in Pinawa was the isolation, especially for those used to a more urban setting. Pinawa did not provide the shopping or cultural opportunities that many desired. This was alleviated by regular trips into Winnipeg for the services and activities not available in Pinawa. This situation remains an issue for residents of Pinawa to this day.

Work at WNRE was an exciting daily experience for most of us. It provided ample opportunities for training and travel. Employees demonstrated the true spirit of working as a team for the common good. I can truly say that I looked forward to the challenges and excitement of going to work each day. There were many memorable experiences while working at WNRE. Some of these included the PIE of the many novel and experimental reactor fuels and fuel designs, major fuel failure events, and power reactor fuel channel fretting campaigns. PIE was even conducted on fuel samples sent to us from the Three Mile Island Reactor accident. Another notable project was the Direct Use of Pressurized Water Reactor Fuel in CANDUs (DUPIC) Project. This involved recycling used Pressurized Water Reactor (PWR) fuel and by a process of oxidation and reduction, this used fuel was re-fabricated into new fuel pellets and assembled into elements for re-irradiation ultimately in a CANDU reactor. All of this performed in the Hot Cells. This was a joint project with the Korean Atomic Energy Research Institute (KAERI) and ultimately resulted in a trip to Korea for me, to present papers on facilities and research programs at WL. One particularly memorable assignment was the work we did to identify the remnants of the Russian satellite Cosmos 954 that broke up during re-entry over northern Canada. Co-incidentally this satellite entered Canadian territory in 1978 on my birthday, January 24th. This satellite was powered by a small nuclear reactor and as most of the debris recovered in the north was radioactive, WNRE was asked to conduct the examination of the material recovered in order to identify the health risks to the inhabitants in the area from the fallout and any impact to the environment. Much of this work was performed in the WNRE Hot Cells, all of it under the close watch of the Americans who were very interested in this satellite's nuclear technology.

Work for AECL provided one with the unique opportunity to change jobs or careers while still remaining within the same organization. For me it was many years involved with the Hot Cells, followed by appointments in Marketing and Business Development (which included a 3 month assignment to AECL's office in Rockville Maryland). This was followed by an attachment to the CRNL Reactor Fuels Branch (although I remained at the WL) where I was responsible for the examination and report preparation of fuel irradiation experiments. My continuous career at WL ended with the formation of the initial team to develop plans and methodology for the decommissioning of the Whiteshell site. Actually this was not a highlight but rather a sad time in the history of the Whiteshell Laboratories. The work conducted at this site over the years was leading edge in many disciplines and to experience its closure was profound. It reminded me of the Avro Arrow incident many years earlier.

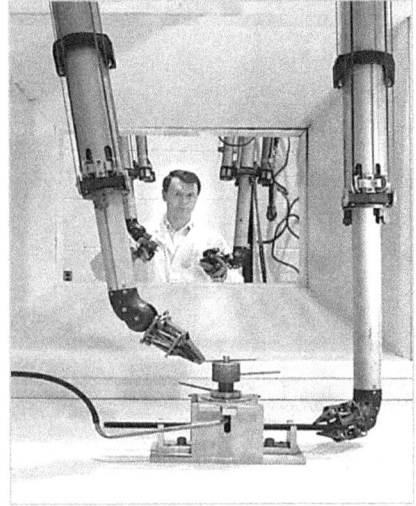

Marvin Ryz at a mock-up of a hot cell in 1967

While I did retire several years ago, I was enticed to return to assist with various facility upgrading and refurbishment projects both here and at CRNL. Currently I continue to work to assist in WL decommissioning projects.

Mike Wright, Henry Marciniak, Mitch Ohta and Mike Tomlinson

125

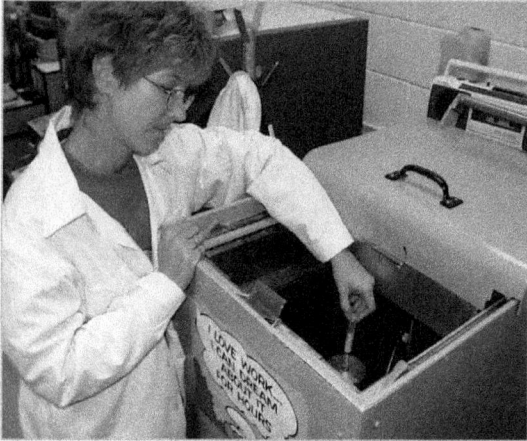

Brenda Wilgosh

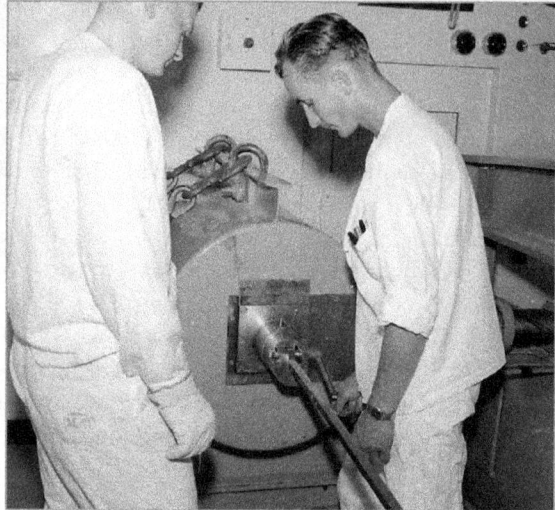

Rodney Zink and Len Woodworth

Glen Brincheski

Chris Saunders, Bob Porth, Tim Shewchuk, Sampat Sridhar,
Nat Fenton and Syd Jones at the Waste Immobilization
Process Experiment

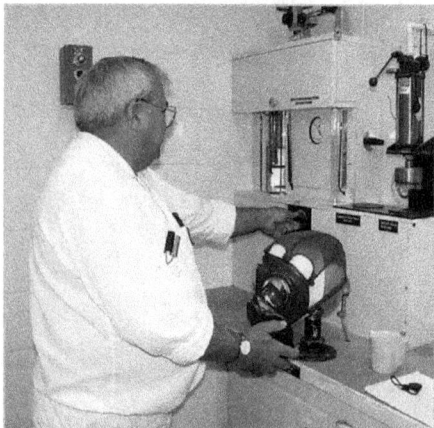

Andy Moreau

Dennis Cann

Blair Skinner

Lori Graham

Margaret Hocking

Siegrun Meyer

Kent Truss

Peter Taylor, Norm Sagert, Neil Miller, Robert Lemire, Derek
Owen and Sham Sunder

David Jobe and Clara Wren

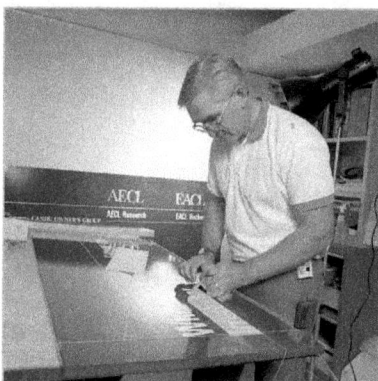

Bob Payne

Randy Herman

David Haworth

David Haworth was born in Birmingham, England, where his father was Professor of
Chemistry at the University and later to be Sir Norman and a Nobel Laureate in
Chemistry. Because of the threat of heavy bombardment of Birmingham in the
Second World War David was evacuated to Canada and joined the household of Sir
Frederick Banting in Toronto for four years. He returned to Britain in 1944 and
continued his education at the University of Oxford. He graduated with a BA in
chemistry which was followed by an MSc at the University of London.

David then went into industrial chemical research with several firms both in Britain
and Canada and in 1971 he was appointed Information Officer at Atomic Energy of
Canada in Pinawa, Manitoba, where he remained until his retirement. David had many interests apart from
chemistry. He taught himself Russian and was an avid reader on many topics.

Len Simpson, Colin Allan and Stu Iverson at "History of Whiteshell
Labs" Lectures – 2013

Radiation and Industrial Safety Branch: Ishwar Garg, Stan Pleskach, May Heinrich, Wayne Sieg,
Carolyn Kirk, Chuck Murphy, George Schultz, Shirley Bell, Ray Lambert and Kent Truss

Whiteshell Memories – My Introduction to Nuclear Research
By Chris Saunders

I came to Whiteshell Laboratories from the University of New Brunswick in 1982. I remember as I drove into Pinawa for the first time that two things stood out; the colour of the soil (red was the colour of the east coast) and the lack of churches in town. Peter Sargent and Alf Wikjord gave me my first job at AECL and it has led to some great experiences and memories. Frank Walton and Nate Fenton were my early office mates.

I initially worked in waste management research, developing methods to vitrify nuclear waste in glass. The team included Sampat Sridhar, Tim Shewchuk, Bob Porth and Sid Jones, with help from many others around Whiteshell. The technology had advanced internationally to the stage that remotely operated facilities were being designed and constructed to demonstrate the process and to vitrify actual waste. AECL wanted to keep up with the technology so we built our own facility in the Building 300 High Bay. We focused our work on the design, building and operation of a rotospray calciner, to convert the nitric acid waste solutions to a dry calcined oxide/nitrate mixture, and a Joule-heated electric melter to produce a waste-loaded borosilicate glass from the calcined wastes plus glass-frit additives. This equipment formed the heart of the Waste Immobilization Process Experiment

Chris Saunders

(WIPE), in which non-radioactive waste solutions were incorporated a borosilicate glass.

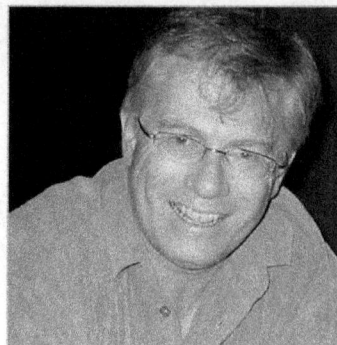

The calciner operated by spraying the feed liquids onto a heated rotating drum. Volatiles, particulates and gases were retained by an off-gas system. The calciner operated with a throughput equivalent to 4 kg/h of waste. The electric melter used start-up heaters to raise the temperature of a waste/glass frit mixture to about 800°C, when the mixture became sufficiently conducting to act as its own resistance heater. Melting at 1100 to 1200°C was achieved by joule heating, using Inconel 690 electrodes. When the melter was filled, the glass flowed over a weir and into stainless-steel containers. The glass was then annealed at 450 to 500°C before cooling. Glass was produced at about 10 kg/h.

In 1987, I joined a group called the Radiation Application Research Branch (RARB), headed by Dr. Stu Iverson. The group developed commercial opportunities for the new linear accelerators AECL began to manufacture. We worked on so many different products and processes, including sterilizing medical products and bones, curing composite parts, destroying hazardous chemicals, food irradiation and studies on the effects of irradiation on the properties of plastics and other organic compounds. Dr. Iverson, Dr. Ajit Singh and Dr. Jos Borsa were important leaders within the group and they helped guide the team to many successes. This work led directly to Acsion and another whole career outside of AECL.

In 1998, nine employees of AECL joined to invest in Acsion Industries. I was proud to be Acsion's first President and board Chairman. Acsion planned to continue the activities of the RARB and to convert our research into new businesses based in Pinawa. Since 1998, we have provided over 250 years of employment, help start several companies, including Acetek Composites, Prairie Isotope Production Enterprise (PIPE), and the Composite Innovation Centre (CIC), and worked on important Canadian projects such as finding new ways to manufacture medical isotopes and working to safely decommission nuclear sites around Canada.

AECL was a great company to start my career. We had world class scientists and engineers that made coming to work an adventure every day. I learned a lot and I continue to be proud of my time there.

Biomass Section of Radiation Applications Research Branch – John Merritt, Terry Stepanik, MiyokoTateishi, Jos Borsa, Srin Rajagopal, Don Ewing, and Richard Whitehouse

George Scharer

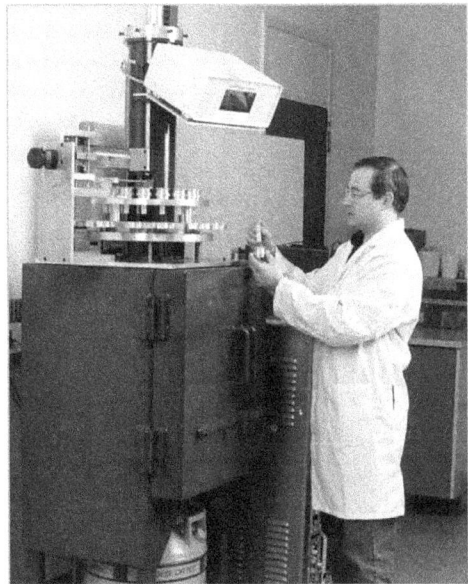

Ernie Bialas

Whiteshell's Originality Circle - Back Row: Ralph Green, Ian Gauld, Tony Wiewel, Phil Davis, Steve Mihok, Mike Tomlinson, Steve Sheppard, Metro Dmytriw; Front Row: Alan Gibson, Keith Mayoh, Janet Dugle

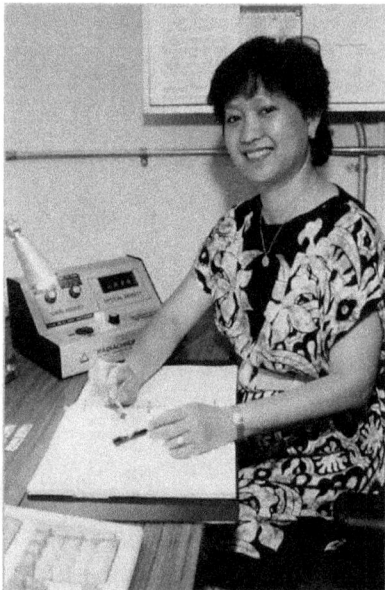

Minda Chung

Vince Lopata

John Stefaniuk, Tom Lamb, Stan Hatcher and Dave Beeching at contract signing.

Rao Puttagunta

Neil Campbell, Russ Webb, Orville Acres, Dave Dugle, Bill Chelack, Jan Dugle, Jos Borsa, Jim Raleigh, and Craig Campbell

Final Shut Down of WR-1 – May, 1985: Left to Right – Barry Hood, Larry Meyer, Ron Feicho, Greg Poole (at Control Console), George Snider, Roy Styles, Ad Zerbin, Mike Wright, George Sharer, Ron Wiggins (behind George), and Wally Dobush

Harry Johnson, Darren Praznik, Colin Allan, Jeff Bishop

Peter Hayward, Keith Harvey and Chris Saunders

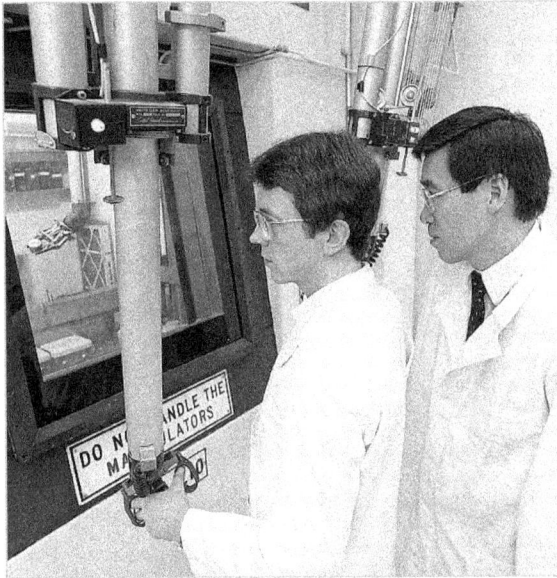
Andy Kerr and Dennis Chen

Walter Harrison and Heiki Tamm inspecting
hydrogen containment vessel

AECL Safety Review Committee - R.B. Lyon, D. Hamel, K. Meek, B.P. Brady, R.V.
Osborne, J.J. Lipsett, J.A. Bond, D.J. Richards, R.J. Hawkins, J.M. Miller, J.W. Logie, R.T.
Jones, R.S. Graham and D.H. Charlesworth

URL 255 m Ceremony – 1985: Scott Ager (1), Glen Snider (2), Doug Peters (3), Gary Simmons (4), Jack Ayotte (5), John Kerr (6), Doug Black (7), Charlie Hirst (8), Fred McPhee (9), Greg Kuzyk (10), Rick Backer (11), John Berry (12), Jennifer Geddes (13), Olivia Tufford (14), Gig Hiltz (15), Ole Vik (16), Don Solberg (17), Don Murray (18), Tim Bryson (19), Jerry Zechel (20, Paul Thompson (21), Cliff Davison (22), Harry Backer (23), Richard Everitt (24), Glen Karklin (25), Earl Masarsky (26), Don Daymond (27), Peter Lang (28), Ron Larocque (29), Tom Boyle (30)

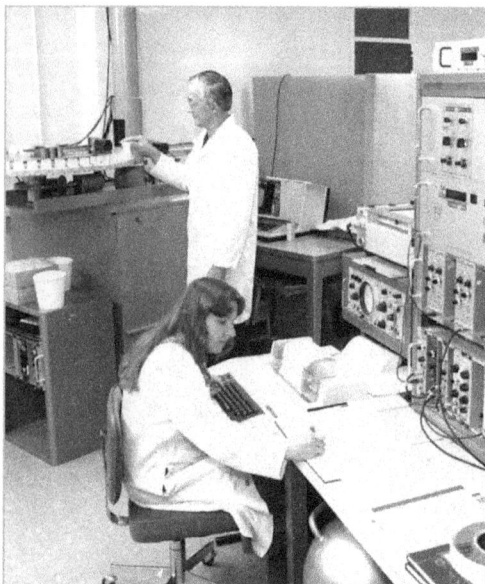

Bill Boivin and Bev Ford

Ken Dormuth

URL Staff: Dick Winchester, Tom Boyle, Derek Martin, Dave Woodcock, Glen Snider, Tim Bryson, Cliff Kohle, Don Solberg, Unkown, Larry Rolleston, Greg Kuzyk, Unknown, John Kerr, Peter Roach, Unknown,Unknown, Gary Simmons,Unknown, Paul Thompson, Dwayne Onagi, Doug Peters, Ray Fillion, Shawn Keith, Don Daymond, Gary Grant, Dan Good, Dwayne Kroll, Bruce Andrews, Wendy Payette, Charlie Hirst, Jason Martino, Unknown, Joyce Mitchell, Ed Dzik, Vern Steiner, Alden Bushman, Richard Everitt, Ed Wuschke, Sam Simcoe, Hugh Spinney, Florence Schimekel, Marlene Payette, Darlene Keith, Gig Hiltz, Tara Kohle, Jill Kaatz

Front Row - David Hnatiw, Hugh Spinney, Dave Woodcock; Second Row – Peter Roach, Vern Steiner, Shawn Keith, Glen Snider, Bob Hampshire, Paul Thompson

Commercial Office: Glen McCrank, Anita Schewe, Jeff Bishop, Candy Bishop, Sandy McDowall, Joyce Tabe, Contractor, Marilyn Bratty, Pat McCooeye, Fred Doern, Don McLean, Bill Seddon

AECL/AGRA Agreement Signing in 1992 – Front Row: Dan deVerteuil, Colin Allan, Two AGRA Employees; Back Row: Bruce McClinton, Frank McDonnall, Ken Dormuth, Anita Schewe, Jeff Bishop, Don McLean

Cliff Zarecki, Bruce Lange, Dennis Cann, Fred Doern and Bill Kupferschmidt

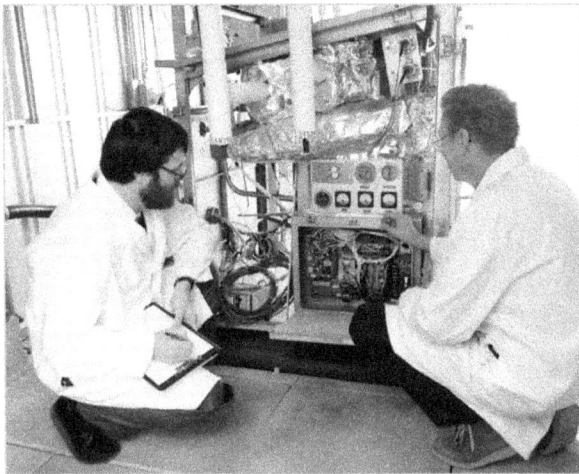

Gary Haacke and Jay Hawton

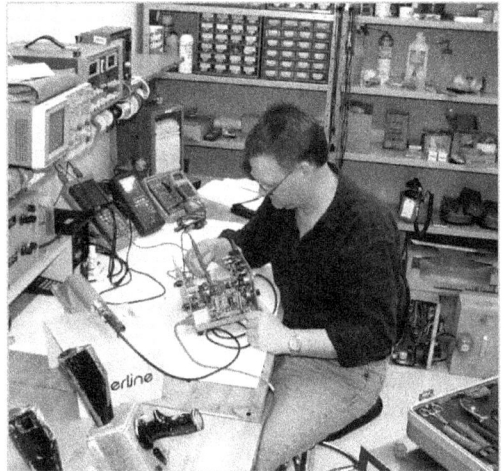

Kim Reimer

Ed Bueckert

Whiteshell Technical Review Group: Front Row – Michael Stephens, Bruce Goodwin, Jim Walker, Terry Andres, Robert Lemire, Ted Melnyk; Back Row – Frank Garisto, Nava Garisto, Reto Zach, Peter Vilks, Phil Davis, Brian Amiro, Unknown, Lyn Ewing

Myrna Tiede, Grant Bailey and Joanne Sieg

Arlene Boivin, Dennis Chen, Wendy Chen and Bill Boivin

Bryan Thomas, Janet Loisel-Sitar and Bill Dewitt

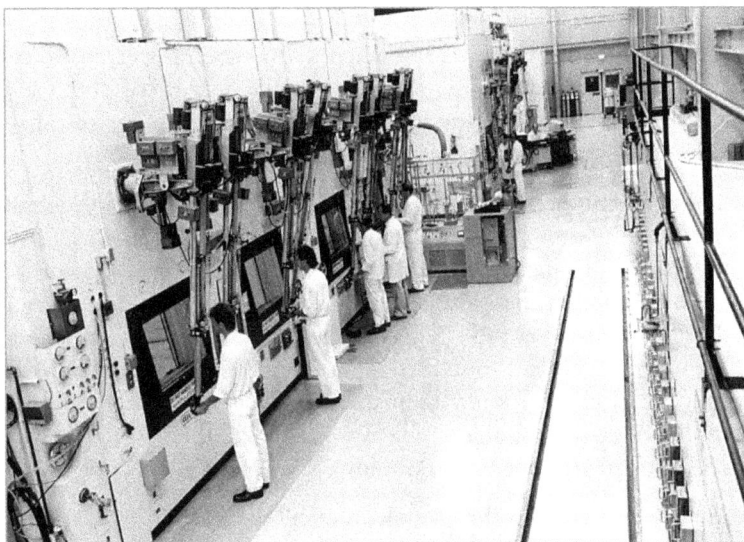

Working in the hot cell facility

Projects Delivery Division, Whiteshell Laboratories – Front Row: Sherri Kuhl, Don Howlett, Alfie Voth, Janice Hawkins, Shannon Stefaniuk, Trish Erickson, Sandi Matheson, Adrian Petrea; Middle Row: Dan Gagnon, Joanne Sieg, Nikki Cure, Enrique Escasura, Janet Stefaniuk, Angela Redmond, Randall Ridgeway, Alex Lafreniere, Lori Bachman, Jayson Cyncora, Shawn Keith; Back Row: Peng Chang, Iouri Minonkov, Jeff Miller, Bruce Martini, Shamsul Alam, Carl Sabanski, Doug May, Terry Stepanik, Brian Wilcox, Grant Koroll, Michael Baader, Mark Kaltenberger, Bill Cunliffe, James Barrios

Whiteshell Brat Perspective
By Pam Meeker

I had just turned seven years old the summer of 1964 when my family transferred from Deep River Ontario to Pinawa. Our Dad was an operator at CRL and was part of a crew of about a dozen that took the transfer out here. He came mostly for the rumours of good hunting and fishing and he wasn't disappointed. Our freezer was always full of venison, ducks, geese and fish.

The reactor was built and went critical in November of 1965. Our dad was in the control room at start up and while I didn't realize it at the time, he was an important part of the history of nuclear power in this country. He was a shift worker for most of his life and he truly loved WR-1. I worked at WL in 1976 and was given the grand tour by my father one slow day at work. At one point in the tour, in the reactor hall, he stopped, stood dead still and looked at me. "Can you feel it?" he asked. I remember standing there with him and I did feel the energy underneath our feet like a living thing. I remember thinking this is what standing on a volcano must feel like. I knew then why he loved working there. He held the reactor in some sort of reverence, like a respected elder. He was proud of the work he did, the way he did it and we were proud of him.

Pam Meeker

I remember the first time we drove down the "9 mile stretch" thinking that the road was similar to the back roads around Deep River and Chalk River that always lead us to a river or lake, so that was fine with me. Then we entered the town and all the roads were still gravel. We found out that our new house wasn't ready for us so we lived in a row house on McGregor for a few weeks. As kids, my sisters and I didn't really comprehend that this was a brand new town, built just for us.

It was a town full of young families and single adults; no grandparents or extended families because the majority of us had moved from other provinces or countries. We were exposed to many languages and accents: British, French, Irish, Scottish, American, European, East Indian, German and the Meekers brought our own Ottawa Valley accent.

We met new children that had lived in Deep River like us but went to another school because they were Catholic and at the time, Catholics and Protestants went to separate schools in Ontario.

There was another separation in Deep River aside from religion, but as a kid I was totally oblivious to the hierarchy of a company town. Apparently in Deep River, the "professionals" didn't socialize frequently with the "hourly rate" employees. There was an invisible but clearly defined line drawn between the two layers of employees at CRL. While I am sure it was evident in Whiteshell, the line was not as defined, probably due to the fact that we were all "imports" and life in general was different in the Prairies compared to our "cultured" lifestyle in the East.

We kids romped and played around our new digs in our Manitoba wardrobes with high rubber boots on our feet to wade through the gumbo and nylon stockings on our heads to combat the bugs. The new houses were gifted with 6 feet of sod around the perimeter and the rest was mud. The next summer the top soil started being trucked in so that we could plant grass all over the yard. I've often thought of our poor mother constantly trying to keep the new waxed hardwood floors clean with 6 pairs of boots trucking in mud until the snow flew that fall. The gumbo was so thick that it wasn't unusual to see a pair of empty boots stuck in the mud waiting for Dad to come home and pull the boots out of the muck.

Growing up in Pinawa was a gift that WL made possible. We had the bush just outside our door and we took advantage of every opportunity to explore, build forts, play in the streets and ride our bikes everywhere. We played, swam, skated, boated, skipped rope and tobogganed until the street lights came on. That was our signal to head home. There were more than 50 kids on our crescent at one point and we would have great games of baseball, red rover, tag and hide and seek. Street hockey was played summer or winter with us dispersing like ants when someone yelled "car!"

We were blissfully unaware of the work being carried on at "the plant". In the beginning, I thought they grew vegetables out there because everyone always talked about the plant. We didn't know if we were playing with the boss's kids or if our parents were the bosses. We just knew that Monday to Friday the buses rolled up and down the roads in the morning, collecting the folks who worked "gentlemen's hours" and dispensing them in the same manner at the end of the day. My dad worked irregular hours, so he didn't ride the bus, but the kids whose parents did used to run and greet their parents as they alighted from the big blue and white buses.

Shelley, Pam, Pat, and Shannon Meeker

Because Pinawa was built from nothing, we had a specific date to celebrate the town's birthday. Every year folks would gather one weekend in July for the big parade, beach activities, fire truck rides, canoe races, dances, and best of all, the fireworks. For the parade, we used to decorate our bikes and every kid in town would make their way down the main drag to the delight of the spectators lining Burrows Road. It was a big deal. It still is and I love that. High school classes still use that weekend to celebrate reunions and families gather from afar to enjoy the festivities.

High school graduations used to be attended by the whole town. It didn't matter if you were related to one of the grads; we just felt like a big extended family holding true to the phrase "It takes a village to raise a child".

Pinawa residents brought a wide range of social activities and interests from their previous lives. Our social clubs included drama, choir, band, art, philanthropists, card players, marksmen, sports, writers, crafters, auxiliaries, etc.

Now I find myself working here at "the plant" and I consider myself lucky to work in a building that houses our history. We have a large photo collection that covers both the plant site and town activities over the decades. I administer engineering records, legacy and current, so I am more aware of the work being done here over the years than most. I have always been proud to come from a family in the nuclear industry. I have defended and boasted about it for most of my life.

I believe nuclear energy is the only future we have that will not impact our environment as negatively as fossil fuels. It might not be in my lifetime, but I hope my children and their children will live in a world with less waste and greenhouse gases being produced and we all know what the answer to that is.

Analytical Science Branch – 1997: Back Row – Sol Sawchuk, Dennis Chen, Terry Howe, Andy Kerr, Willie Dueck, Brian Wilcox, Ela Rochon, Ken Wazney, Grant Delaney, Gordon Burton, Clarence Musick, Steve Liblong, Mike Attas, Mike Bratty, Ernie Bialas, Rick Katz, Rich Hamon, Gary Young; Middle Row – Linda Brown, Andy Gerwing, Barb Sanipelli, Randy Herman, Elisabeth Nguyen; Front Row – Mary-Ann Cribbs, Merle Brown, Dana Joseph, Noemi Chauqui-Offermanns

Dwayne Onagi

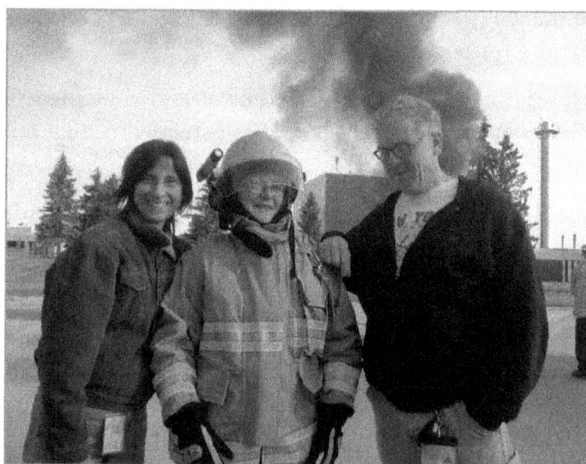

Martha Laverock, Karen Ross and Trefor Jenkins at fire training

Lorne Hachkowski and Andy Gerwing

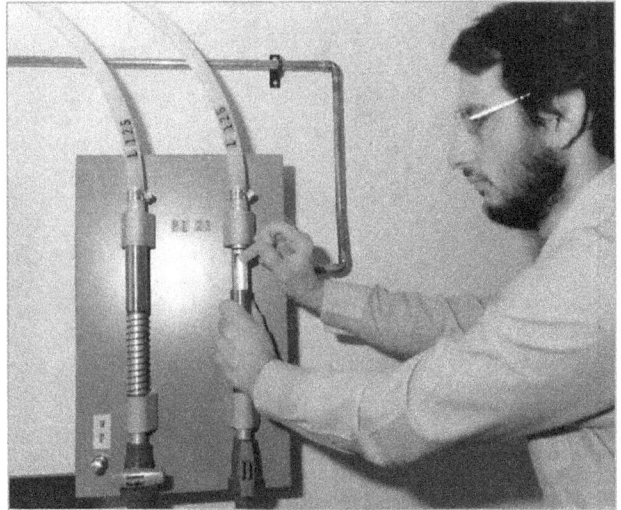

Mike Attas operating the Super-Rabbit NAA facility

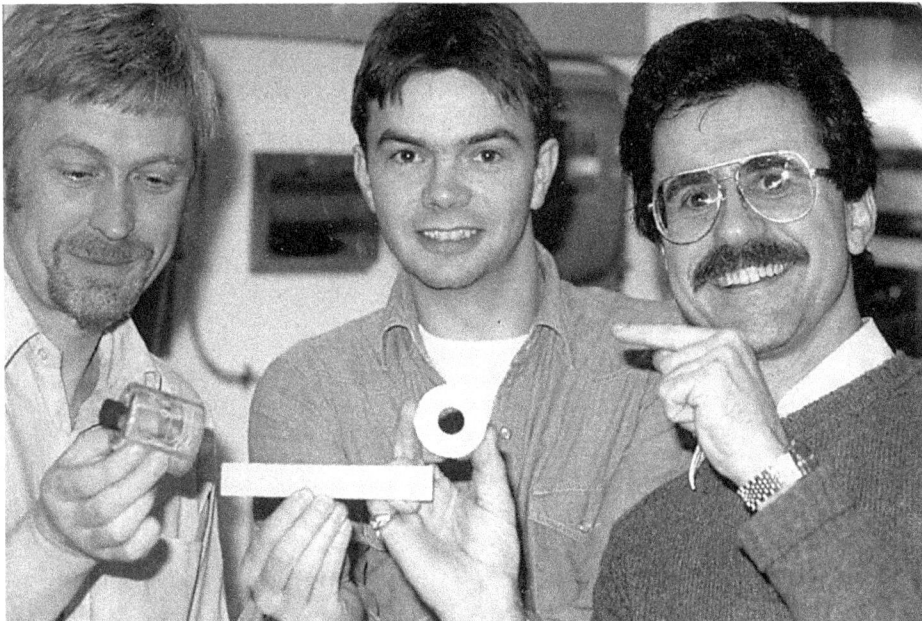

Peter Hayward, Ian George and John Tait

Ten Year Safety Award Presentation: Front Row – Bill Sitar, Bill Klapprat, Gerry Sachvie, Stan Kekish, Dale Lidfors, Phil Clarke, Bill Hyrishko; Back Row – Irvin Hemminger, Gerald Krawchuk, Lloyd Voss, Ron Radons, Doug Platford, Glen Graham, Gerrit Teirpstra, Peter Honnoff

Staff and management opening the SMAG Facility: Jeff Harding, Vernon Pommer, Iouri Minenkov, Kevin Rogers, Joan Miller, Glen MacLean, Randall Swartz, Shawn Keith, Dwayne Onagi, Russ Mellor, Ray Warenko, Doug May, Greg Short, Gary Rollins, Matt Balness, Curtis Graham.

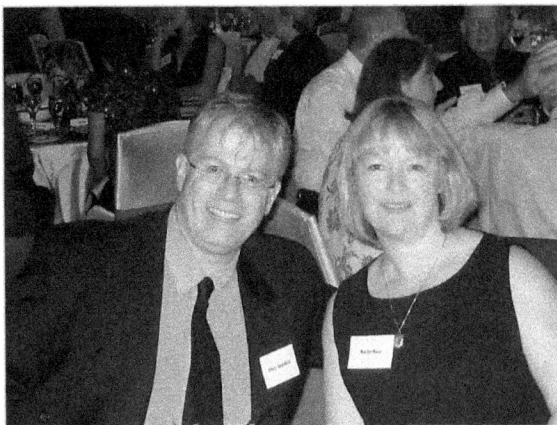

Chris Saunders and Karen Ross at an AECL
anniversary celebration.

.   Garry Schellenberg, Jerry Nayler and Walt
Litvinsky

John Barnard touring Bob McDonald and friends

Everett Dobbin

Hammy Carswell

Dave Smith, Bud Kelly, Ralph Danbert, Don Price, Jerry Nayler. Dale Lidfors and
Betty Hembroff

Design and Projects Branch Personnel from the Past

Cathy Abraham	Tom Fox	Bud Kelly	Eric Payne	Dave Smith
Charla Aitken	Art Gauthier	Dick Keltie	Doug Peters	Ray Sochaski
Jim Aitkens	Lynn Gauthier	Dennis Klepatz	Don Price	Hans Sohn
Bill Ayres	Dick Gillert	Dale Lidfors	Mike Pronozuk	Mary Spitz
Bill Baker	Denis Godin	Walter Litvinsky	Jim Putnam	Ingrid Stark
Doug Benton	Jim Gold	Henry Marciniak	Larry Ramsay	John Stermscheg
Kevin Borgford	Julie Halley	Ron Mazur	Jack Reimer	Garry Stokes
George Bowhay	Jim Harding	Linda McCallum	Jack Remington	Stephanie Suchar
Dennis Byskal	Jeff Harding	Des McCormac	Lynn Reykdal	Dean Sullivan
Hammy Carswell	Deb Hartrick	Sandy McDowall	Emile Robert	Dennis Sutherland
Marv Cooper	Betty Hembroff	Margret McEwan	Diane Robert	Lorne Swanson
Ralph Danbert	Betty Hemminger	Ken Meek	Carol Roberts	Tom Tabe
Vern Decterow	Len Hiebert	Wayne Merlin	Arie Rylaarsdam	Paul Tweitmeyer
Stan Dubeck	Fred Hill	Fred Milke	Carl Sabanski	Ole Vik
Ed Ellila	Mickey Hoban	John Montgomery	Erwin Schatzlein	Sue Wallace
Richard Farquhar	Lynda Jackson	Del Nelson	Garry Schellenberg	John Westdal
Sandy Fay	Ed Jacobs	Roy Nickels	Herb Schmitz	Jim Wright
Nat Fenton	Walter Jaster	Mitch Ohta	Ken Scott	Jerry Wyshnowsky
Harold Fisher	Dieter Jung	Peter Orlick	Gary Simmons	Gertie Zieske
Dwayne Fittner	Ken Kamachi	Don Patterson	Charles Simons	

Reactor Operations Branch Personnel from the Past

Harry Backer	Blake Cutting	John Kerr	Bernie Pannell	Art Summach
Roy Barnsdale	Steve Daley	Cliff Kohle	Vic Parrot	Tom Tabe
Bun Baxter	Rick Day	Ben Kollinger	Gerry Plunkett	Del Tegart
Harold Bender	Wally Dobush	Jake Kruger	Vic Popple	Doug Taylor (1)
Bill Berry	Brent Donnelly	Fred Legiehn	Ian Prysnyk	Doug Taylor (2)
Mike Berry	Mickey Donnelly	Ken Malyon	Lloyd Rattai	Don Tirschmann
Jim Biggs	Bob Ennis	Gilles Marion	Ross Reid	Roy Ticknor
Dennis Bilinsky	Nat Fenton	Jim Maxwell	Vic Reschke	Pat Tighe
Jack Blacher	Ron Feicho	Vinny McCarthy	Bennie Richter	Brian Tracy
Thor Borgford	Tom Fox	Sandy McDowall	Alex Robertson	Dave Tymko
Evan Bowles	Larry Gauthier	Joe McMann	Gord Robinson	Grant Unsworth
Tom Boyle	Bernie Gordon	Dick Meeker	Ian Ross	Ward Wallace
Bill Brooke	Dennis Graham	Trevor Mellors	Phil Roy	Gerry Walters
Warner Brown	Eric Graham	Larry Meyer	Carl Sabanski	Terry Wardrop
Bob Bruneau	Ken Gray	Jack Middleton	Gary Schwartz	Ray Weber
Don Bruneau	Keith Greenfield	Pat Mills	Wayne Sieg	Wayne Weseen
Louie Bruneau	Ron Gurkie	Al Nelson	George Sharer	Ron Wiggins
Mark Bruneau	Bob Hampshire	Gaylord Newman	Gary Simmons	Don Young
Alden Bushman	Don Hatch	Bob Noble	George Snider	Cliff Zarecki
Wilf Campbell	Al Hawkes	Ron Oberick	Glen Snider	Bill Zuk
Dave Carter	Barry Hood	Mitch Ohta	Klaus Spitz	
Larry Chura	Stu Jansson	Frank Oravec	Wayne Stanley	
Jim Clarke	Al Jarvis	Albert Palson	Vern Steiner	

WR-1 Control Room

Maintenance Branch Personnel from the Past

Ted Anderstedt
Ray Andrews
Bill Ayres
Bob Barks
Kerry Beauchamp
Les Bell
Brian Bilinsky
Ron Bratty
Brian Broulette
Garry Buchanan
Ed Bueckert
Roland Cadoreth
Don Campbell
Ed Campbell
Don Carrier
George Champagne
Lloyd Colborne
Brian Corbett
Ray David
Harley Davidson
Bob Desjardins
Bill Dereski
Rheal Desbois
Everett Dobbin
Hank Dowse
Ron Drabyk
Stella Dreger
Bill Dunford
Tom Dunlop
K. Dykstra
Ed Ellila

Russel Fenning
Ray Fillion
Al Futcher
Philip Gauthier
George Gibson
Howard Gilmore
Cass Gladys
Audrey Graham
Glen Graham
Harold Graham
Ed Grant
Gary Grant
Peter Grzegorzewski
Felix Gryseels
Jim Hadley
Dennis Halliday
John Hannon
Liz Hannon
Peter Hannoff
Bill Hart
Harry Haugen
Kirk Haugen
Irving Hemminger
Ed Hoard
Milt Holochuk
Len Horn
Barry Ivison
Kevin Jackson
Marty Jacobs
Joe Kaston
Wilfred Keith

Tom Kearns
Alvin Klapprat
Geoff Knight
Bill Kohlman
Joe Kostiuk
Ky Kubota
Henry Kuehl
Gerry Lange
Joe Lauret
Bert Lavesque
Dale Lidfors
Bill Litke
Mark Litvinsky
Ed Locke
Ken Lodge
Roy Lucas
Ron MacLean
Harold Malkoske
Jerry Martino
Sandy Mathews
Murry Matiowsky
Gloria McAuley
Ross McConnachie
Bill MacDonald
Gord McDowall
A.C. McKay
Dale McKay
Keith McKentyre
Ed McKerrell
Don Melsness
John Michelfeit

Jack Middleton
Jim Mitchell
Ross Mitchell
Brian Morash
Bob Morris
Gord Murray
Ernie Muzychka
Lou Nagy
Don Nerbas
Larry Novakowski
Lillian Novakowski
Ron Oberick
Mike Ouimet
Roger Parent
Dennis Penner
Don Pihulak
Steve Pihulak
Doug Platford
Jim Putnam
Ron Radons
Bob Randell
John Rankin
Eleanor Reimer
Herb Richter
Alvin Rueckert
R. Ruymar
Mary Ryz
Richard Schultz
Charles Scott
Larry Shorrock
Dennis Smith

Harry Smith
George Snider
Ted Stapleton
Gary Sterling
Wally Stober
Stephanie Suchar
Dennis Thomas
Steve Tiede
Harold Trapp
Garret Trastra
Jack Turner
Grant Unsworth
Elmer Voelpel
Augie Weissig
Ed Weissig
Peter Westhoven
Mike Wayne
Richard Waywood
Ray Wazney
Anthony Wiewel
Al Wilgosh
Dick Willacy
Len Williams
Bill Woodbeck
Joe Zanutto
Ad Zerbin

Stores Group – Building 408: Ken Bjornson, Jim Rogocki, Cas Szajewski, Rosalie Lofstrom, Brent Donnelly, Randy Mamrocha, George Kowalchuk

Engineering, Products and Services Branch – Front Row – Don Saluk, Jerry Krawchuk, Stan Kekish, Peter Grezegorzewski, Richard Farquhar, Paul Sansom, Ben Delannoy, Garry Stokes, Jeff Harding; Back Row – Garry Buchanan, Henry Hubner, Larry Stelko, Ervin Hemminger, Mike Veilleux, Richard Schultz, Ken Urbanski, Edgar Bozak, Bruce Andrews, Kim Reimer, Brian Dyck, Randy Kolesar, Clive Schulz, Gerry Lange

Brent Donnelly surveying a sample

Ron Oberick, Monte Fidler at fire fighter practice

Kevin Larsen, Ken Bilkoski, Cory Wilson, Glen Graham

Maintenance and Materials Branch: Front Row – Glen Wilgosh, Harold Bender, Ed Veroneau, Paul Sansom, Cliff Zarecki, Gary Wallace, Mike Veilleux, Larry Stelko, Len Kroeker; 2nd Row – Garry Buchanan, Fay Dooley, Grace Guttman, Allan Young, Terry McCalder, Len Molinski, Colin Brown, Andy Morneau, Mark Kaltenberger, Phil Clarke, Scotty Wooster, Brian Thomas; 3rd Row – Alf Gesell, Don Kubish, Clive Schultz, Ervin Hemminger, Alvin Rueckert, Horest Sommerfeld, Karen Kovacs, Clarence Newman, Peter Grzegorzewski, Richard Schultz, Bill Sitar; 4th Row – Don Zetaruk, Kevin Borgford, Floyd Carlson, Les Carlson, Kim Reimer, Doug Platford, Les Meyers, Dennis Kabaluk, Gerry Lange, Oswald Meyer, Geoff Knight, Eric Amos, Randy Mamrocha, Lloyd Voss, Larry Novakowski, Wayne Wasylenko

Finance and General Accounting Branch: Irene Hiebert, Cindy Lodge, Dorothy Turner, Carol-Lee Salmon, Gord Markham, Joan Hampshire, Debbie Manchulenco, Ron Loeb, Leslie McClinton, Carol Findlay, Lise Desrocher-Alpers, Susan Marohn, Kathy Hanna, Lori Rodych

Waste Recycling and Disposal – Jeff Strmbiski (above) and Dave Demers (below)

Whiteshell Laboratories

Pinawa – Home of Many Whiteshell Staff

Chapter 8
Appendix

Whiteshell Laboratories – Building Numbers and Description

Building Number	Description
100	WR-1 Reactor
200	Active Liquid Treatment Centre
300	Research and Development Centre
303	Gas Dynamics Research Laboratory
304	Gas Dynamics Research Laboratory
306	Gas Dynamics Equipment Storage
307	Diffusion Flame Facility
312	Steam Generator Storage
400	Engineering and Administration Building
401	Services and Control Centre
402	Health and Safety Building
403	Vehicle Gate House
404	Meteorological Tower
405	Technical Information Centre
406	Cafeteria
408	Stores, Workshop and Garage
409	Active Area Stores
410	Cafeteria Waste Storage
411	Laundry and Decontamination Centre
412	Active Area Workshop
413	Waste Chemical Storage
414	Controlled Area Entrance
415	Material Warehouse
416	Solid Waste Storage
418	Fissionable Material Storage
420	Mobile Equipment Storage Building
422	Outfall Monitoring Station
424	WR-1 Organic Monitor
426	Civil Storage #1 Building
427	Mechanical Shop Storage #1
428	Mechanical Shop Storage #1
429	Civil Storage #2
500	Internal Friction Laboratory
501	Aquatic Toxicology Laboratory
504	Engineering Development and Testing
505	Soils Research Laboratory

509	Civil Utility Building
511-1	Active Waste Storage #1
525	Meteorological Trailer #2
526	Borehole Instrumentation Test Facility
527	Inflammable Liquid Storage Building
530	Internal Friction Laboratory Annex
902	Pumphouse
903	Water Treatment Plant
907	Sewage Lift Station
911	Central Powerhouse
913	Main Substation

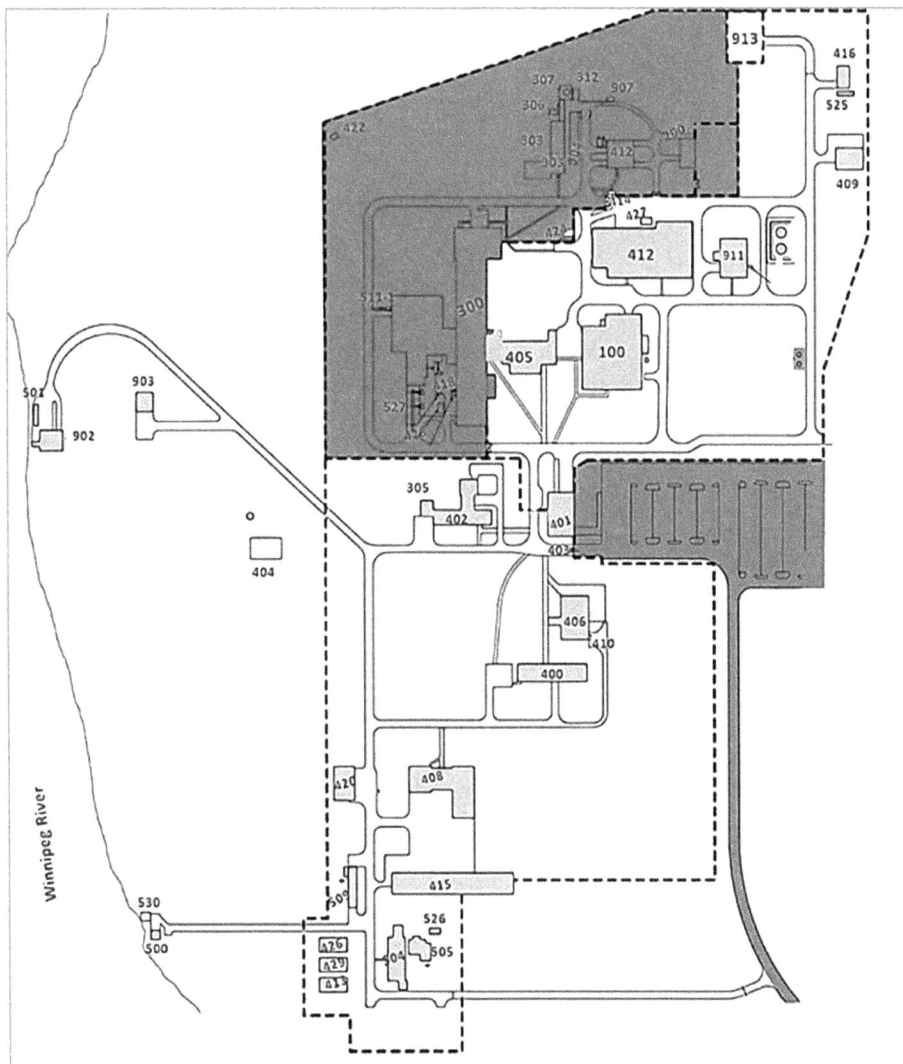

Map of Whiteshell Laboratories

WR-1 Reactor Specifications

GENERAL

Reactor type

Fissionable material: 1.3 - 2.25 wt% U-235
Moderator 99.73% D_2O; Reflector 99.73%D_2O
Coolant - By weight - 70% Monsanto OS84
                      30% Radiolytic Tars

Nominal Reactor Power

60 MWth

Purpose

Engineering test of coolant materials, coolant tube materials, coolant tube design, fuel materials, fuel cladding materials, and fuel design.

Designers / Builders

Reactor Plant - Canadian General Electric
Building - Shawinigan Engineering

Construction Schedule

Start of construction - 1963
Reactor Critical - November 1, 1965
Full Power - January 1966

Mean Neutron Energy in the Core

Thermal

Mean Lifetime of Prompt Neutrons

5.2×10^{-4} sec

Core Parameters

Note: These parameters are dependent on moderator height, core enrichment, fuel burnup, and moderator boron concentration.
For a unit cell (hot, fresh fuel, 2.25 wt% U-235, small calandria tube, 1.0 ppm Boron-10 in moderator):
$e = 1.029$
$n = 1.644$
$p = 0.908$
$f = 0.904$
$L^2 = 127.2$ cm^2
$L^2_s = 125.1$ cm^2
Fast leakage factor = 0.939
Thermal leakage factor = 0.938

Neutron Flux

Thermal average $\sim 5 \times 10^{13}$ n/cm^2 sec
Thermal maximum $\sim 1.5 \times 10^{14}$ n/cm^2 sec

Excess Reactivity Balance

Xenon (100% R.P.) - 27.2 mk
Fuel burnup - 0.4 mk/Full Power Day

Maximum Excess Reactivity Built-in

93 mk (52 in. to 96 in. moderator height)

CORE

Shape and Dimensions

Cylindrical. Maximum Height 96 ins. (244 cm)
Operating Height 89 ins. (226 cm)
Radius 35.36 ins. (89.8 cm)

Channels and Assemblies in Core

Normal coolant circuits: 49

Experimental Loops: 4
Capsule Facility: 1

Self-powered flux detectors: 1

Standard Channel	Material: Ozhenite 0.5 I.D. 3.263 in. (8.29 cm) Wall thickness: 0.125 in. (0.32 cm)
Standard Assembly	Height: 97.5 in. (248 cm) O.D. 3.23 in. (8.2 cm)
Lattice	Hexagonal - Pitch 9.25 in. (23.5 cm)
Critical Mass	(Cold Unpoisoned): 23.3 kg U-235 (19 site core 84.3 in. moderator height) 21.8 kg U-235 (25 site core 59.2 in. moderator height)
Average Specific Power in Fuel	1.1 MW/fuelled site (at full power, 89 in. moderator height) 4.8 kW/cm fuel
Average Power Density of Core	294 W/cm³ of fuel (at full power, 89 in. moderator height)
Blanket Gas	Helium

FUEL

Material	Cast UC slugs, cylindrical sheath.
Cladding	Zr-2.5 Nb
Elements	Sheath O.D. 0.530 in. (1.35 cm) Fuelled length 18.78 in. (47.7 cm) Total length 19.5 in (49.5 cm)
Subassemblies [bundles]	14 elements/bundle - uniformly spaced in a single ring circular lattice. Centre position contains hollow Zr-2.5 Nb support rod.
Assemblies	Length of fuel end ~97.5 in. (248 cm) Length of fuel assembly ~19 ft. (580 cm)
Burnup	Rated average 360 MWh/kgU
Fuel Loading & Unloading System & Procedure	Organic cooled transfer flask, transported by main station crane. Loading and unloading off-power from above core.
Irradiated Fuel Storage	Spent fuel storage - 182 assemblies - fuel stored in organic filled cans - cans stored vertically in fixed configuration in water-filled storage bay. Fuel storage block - 26 storage tubes - fuel can be stored in organic or dry - tubes immersed in water - designed for short-term storage. Canisters - as required (144-222 bundles per canister) - fuel stored dry in seal-welded steel baskets within 0.75 m thick, reinforced concrete canister.

CORE HEAT TRANSFER

Heat Transfer Area	~13,500 cm² per fuel end (assuming radial heat transfer)
Heat Flux on the Fuel Element Surface	Maximum 160 W/cm² Average 75 W/cm² (over length of fuel end)
Film Temperature Drop	~60°C
Maximum Design Fuel Temperature	1100°C
Maximum Design Clad	500°C

Whiteshell Laboratories

Surface Temperature	
Coolant Flow Area	~750 cm²
Channel Velocity of the Coolant	11.5 m/s - nominal velocity in average site
Coolant Mass Flow Rate	750 kg/sec.
Coolant Temperature/ Pressures	Temperature: Inlet 280°C - 400°C Outlet 320°C - 425°C Pressure: Inlet Header 315 psig (2.15 MPa) Outlet Header 160 psig (1.1 MPa)
Provision for Shutdown Heat Removal	Pressurizing pumps provide flow through standby coolers.

COOLING SYSTEM

Heat Exchangers	3 circuits - 1 HX/circuit - shell & tube U-bend, single pass, water cooled tubes of seamless steel clad inside with 70/ 30 copper/nickel
Primary Coolant Losses/ Decomposition	Estimated damage rate at 30% high boiler ~4 kg/MWD. Degassing and particulate removal systems operate on continuous basis. Insoluble gases are discharged to the atmosphere.
Safety Features of the Cooling System	Pressure relief valves. Emergency injection tanks pressurized with nitrogen and containing 100,000 lb (45,000 kg) of coolant will release into the inlet header if pressure drops below 150 psig (1.1 MPa). Water can also be injected. Coolant pumps equipped with fly-wheels to give proper coast-down characteristic.
Provision for Detecting Fuel Element Defect	Gross gamma scan on coolant outlet feeders.

CONTROL

Maximum Rate of Reactivity Addition	0.33 mk/sec
Trip Mechanism/Time	Mechanism - moderator dump Reactivity reduction 1 sec after trip signal: 13.8 mk 1.5 sec after trip signal: 25.7 mk
Sensitivity of Automatic Control	3 %
Temperature Coefficients	Moderator: +0.14 mk/°C at 1 ppm B^{10} and 20°C Coolant: -0.03 to +0.01 mk/°C Fuel: -0.045 to -0.025 mk/% Rated Power
Burnable Poison	Boron (93 wt% enriched in B^{10}) in the form of Boric acid dissolved in the moderator
Other Control / Shutdown Provisions	Control and regulation primarily by variation of moderator level. Boron-10 concentration in the moderator can be varied with injection pumps and ion exchange columns providing ~28 mk/ppm B^{10} of reactivity load

REACTOR VESSEL

Form, Materials/ Dimensions   All welded vertical cylinder of ASTM A240 stainless steel, 16.5 ft. (5 m) high, 8 ft. 10 in. (2.7 m) I.D., tapered on top side with dished top and bottom.
Wall thickness 0.5 in. (1.27 cm)
Top head thickness 0.75 in. (1.91 cm)
Bottom head thickness 1.375 in. (3.49 cm)

Design and Working Pressures   Design: 50 psig (3.52 kg/cm²)
Working: 8 psig (0.56 kg/cm²)

REFLECTOR & SHIELDING

Reflector   Heavy Water - 99.73%
Thickness - 17.64 in. (44.8 cm)

Radiation Level Outside Shielding   less than 2 mR/h in all unrestricted areas

Shielding   Side: 5 in. (12.7 cm) steel + 6 in. (15.2 cm) water
7 ft. 6 in. 9229 cm) of 220#/ft. (3.52 gm/cm³) concrete

Top: 22 in. (55.9 cm) steel + 22 in. (55.9 cm) water (shield)
18 in. (45.7 cm) steel + 3 in. (7.6 cm) masonite (deck plate)

Bottom: 22 in. (55.9 cm) steel + 22 in. (55.9 cm) water
Cooling: Water circulation
Max Temp: Side 116°C ; Top 63°C ; Bottom 52°C
Average Cooling Water Temp: 35°C

Reactor Overall Dimensions with Shielding   Approx. 32 ft. (9.75 m) high x 26.5 ft. (8 m) diameter

CONTAINMENT

Containment Type/ Material   Normal Industrial type building with absolute filter protection on ventilation effluent

Surroundings   Locate adjacent to Whiteshell Provincial Park. Total population within 25 mile radius - less than 8,000.
Seismology - Zone 1 on Canada Earthquake Probability Map - only slight damage has been recorded.
No single wind direction >25% frequency.

RESEARCH FACILITIES

Special   Any coolant channel can be isolated from the system and run as an experimental loop. Similarly, any pressure tube can be removed from the reactor and replaced by a tube of different design, provided only that it fits in the calandria tubes.
The coolant system is split into three circuits: thus three organic coolants can be tested simultaneously.

Experimental Loops   Four experimental loops are installed, and there is room for two additional loops.

In-Core Irradiation Sites   Three pneumatic capsule tubes transport samples for irradiation into an outer site in the core. One is connected directly to the Hot Cells, the other to the Radiochemistry Laboratory.

Special fuel hanger rods may be installed in any site in the reactor. Samples up to 1.75 in. (4.5 cm) in diameter can be irradiated in the centre of a fuel assembly.

Three experimental loops are fuelled with an assembly mounted on a large hollow hanger rod. Devices (e.g. creep machines, steam autoclaves, etc.) up to 2.8 in (7.1 cm) in diameter can be operated in a fast neutron flux (>1 MeV) of ~10^{14} n/cm^2 sec.

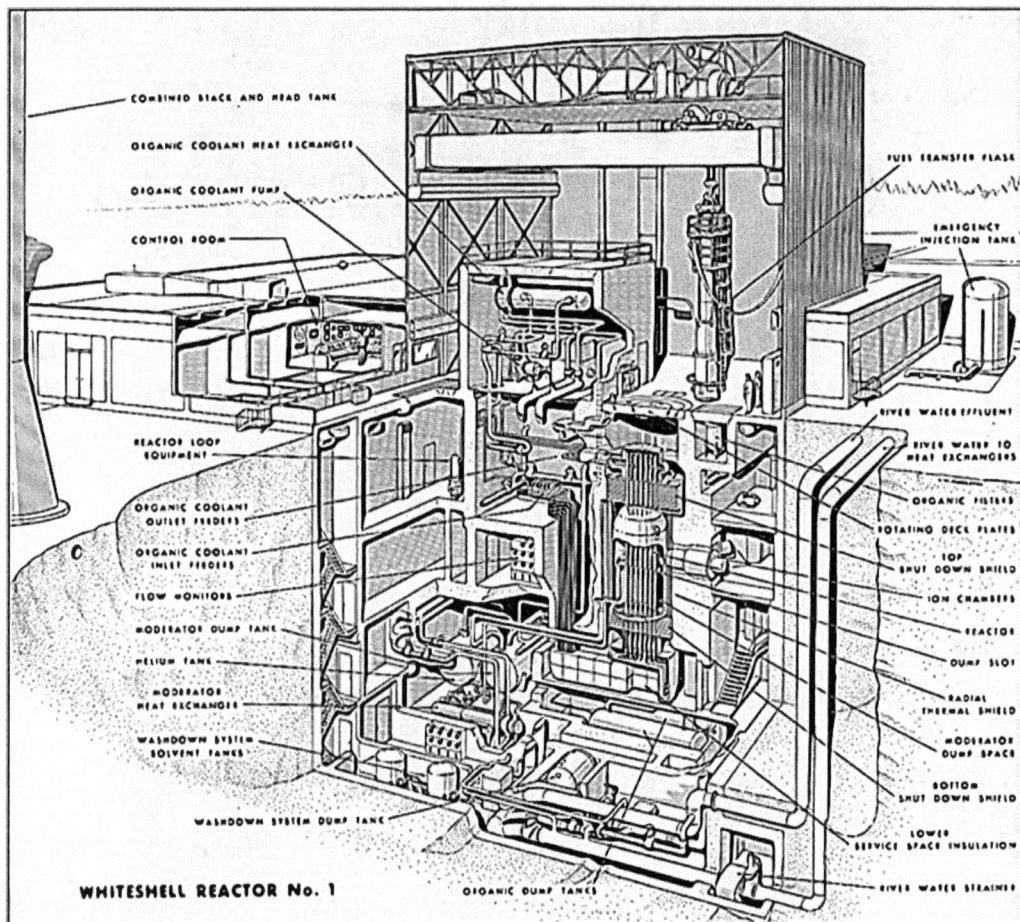

WHITESHELL REACTOR No. 1

Whiteshell Organizational Structure – September 1982

VP & General Manager
R.E. Green

Assistant to VP
R.S. Dixon

Director – Safeguards Development
R.M. Smith

Manager – Commercial Operations
R.O. Sochaski

Environmental Authority
R.W. Pollack

Director – Chemistry and Material Science
M. Tomlinson

- **Materials Science Branch** — R. Dutton
- **Analytical Science Branch** — R.B. Stewart
- **Research Chemistry Branch** — D.F. Torgerson

Director – Waste Management Division
T.E. Rummery

- **Fuel Waste Technology Branch** — K. Nuttall
- **Geochem. & Applied Chem. Branch** — F.P. Sargent
- **Applied Geoscience Branch** — K.W. Dormuth
- **Environ. & Safety Assess Branch** — R.B. Lyon
- **Waste Management Public Affairs Section** — E.R. Frech

Director – Applied Science Division
W.T. Hancox

- **Technical Services Branch** — D.W. Jung
- **Thermalhydraulics Research Branch** — W.T. Hancox*
- **Systems Analysis Branch** — W.C. Harrison
- **Fuel Recycle Branch** — D.R. McLean
- **Materials and Mechanics Branch** — M.B. Wright

Director – Health and Safety Division
J.L. Weeks

- **Environmental Research Branch** — S.L. Iverson
- **Medical Services Branch** — R.J. Hawkins
- **Medical Biophysics Branch** — A. Petkau
- **Rad. & Indus. Safety Branch** — E.J.K. Cowdrey
- **Environ. Authority** — R.W. Pollack

Manager – Eng. Design and Ops Division
S.A. Mayman

- **Design and Project Eng. Branch** — M.M Ohta
- **Reactor Technology Branch** — R.W. Pollock
- **Reactor Operations Branch** — A.R. Robertson

Manager – Maintenance and Construction Div.
J.A. Martino

- **Electrical Instrum. & Power Branch** — A. Zerbin
- **Civil and Mechanical Services Branch** — E.D. Lidfors

Manager Administration Division
D.K. Beeching

- **Personnel & Pinawa Accommodation Services Branch** — H.M. Johnson
- **Public Affairs Branch** — M. Dmytriw
- **Tech. Information Services Branch** — M.O. Luke
- **Protective Services** — T.B. Lamb

Manager Finance
D.W. Murray

- **Purchasing Branch** — B.M. Bjornson
- **Finance Branch** — D.W. Murray*
- **Stores Services** — J.P. Williamson

R & D Administrative Services — J.H. Wright

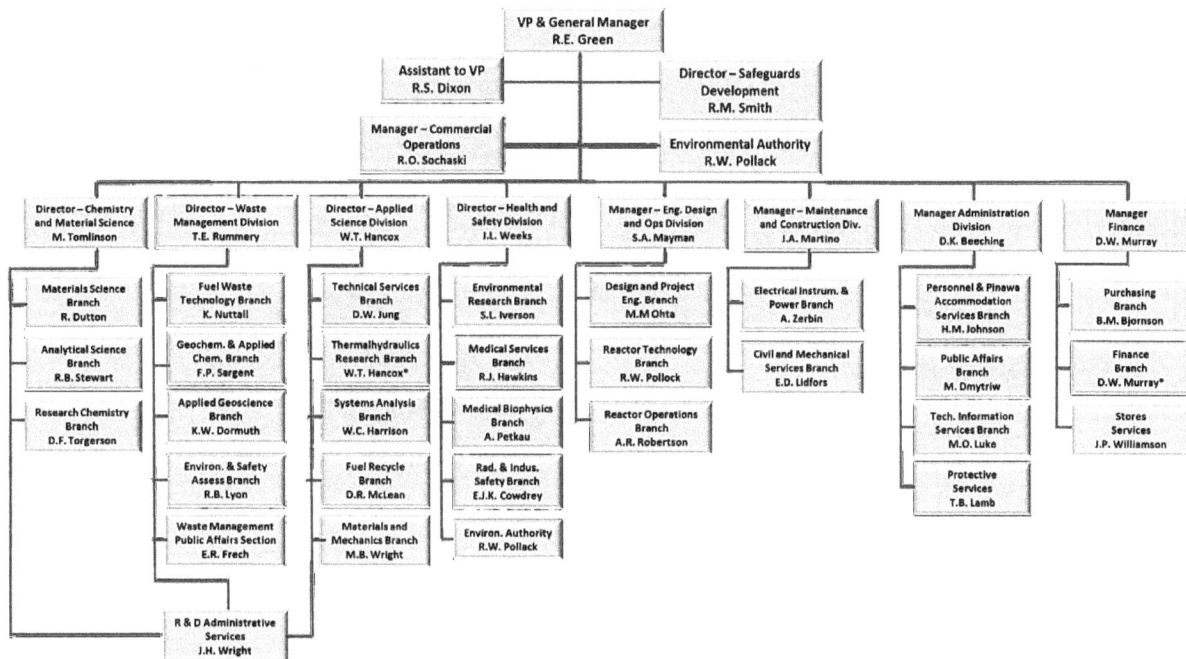

Whiteshell Organizational Structure – September 1982
(* acting position)

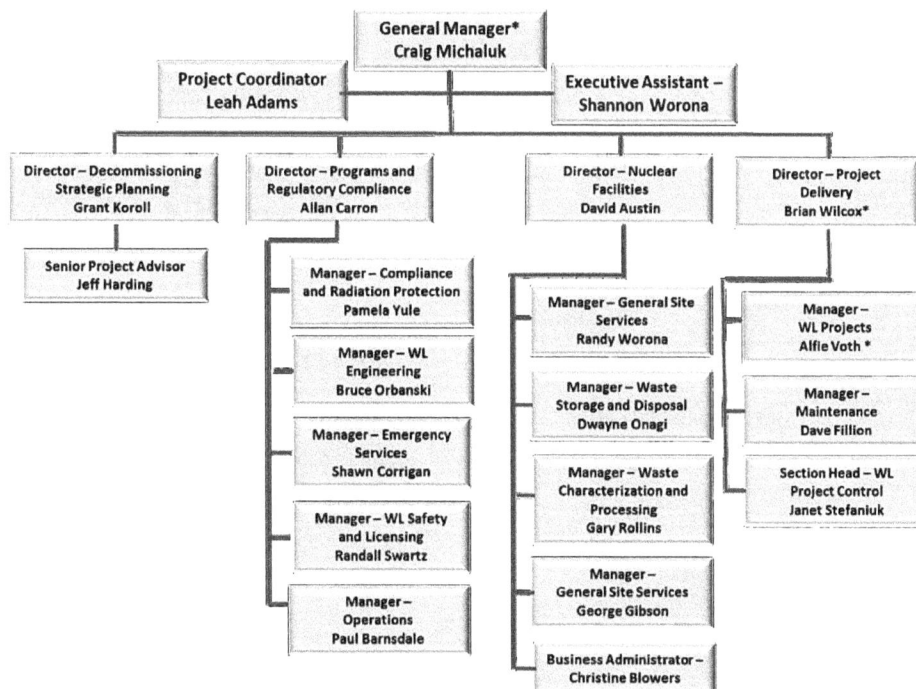

Whiteshell Organizational Structure – November 2014

General Manager*
Craig Michaluk

Project Coordinator
Leah Adams

Executive Assistant –
Shannon Worona

Director – Decommissioning Strategic Planning
Grant Koroll

- **Senior Project Advisor** — Jeff Harding

Director – Programs and Regulatory Compliance
Allan Carron

- **Manager – Compliance and Radiation Protection** — Pamela Yule
- **Manager – WL Engineering** — Bruce Orbanski
- **Manager – Emergency Services** — Shawn Corrigan
- **Manager – WL Safety and Licensing** — Randall Swartz
- **Manager – Operations** — Paul Barnsdale

Director – Nuclear Facilities
David Austin

- **Manager – General Site Services** — Randy Worona
- **Manager – Waste Storage and Disposal** — Dwayne Onagi
- **Manager – Waste Characterization and Processing** — Gary Rollins
- **Manager – General Site Services** — George Gibson
- **Business Administrator –** Christine Blowers

Director – Project Delivery
Brian Wilcox*

- **Manager – WL Projects** — Alfie Voth *
- **Manager – Maintenance** — Dave Fillion
- **Section Head – WL Project Control** — Janet Stefaniuk

Whiteshell Organizational Structure – November 2014
(* acting position)

Whiteshell Staffing

Full Time Employees

1962	27	1989	937
1963	51	1990	955
1964	187	1991	951
1965	321	1992	923
1966	424	1993	901
1967	567	1994	816
1968	678	1995	739
1969	754	1996	703
1970	780	1997	641
1971	784	1998	582
1972	791	1999	313
1973	785	2000	296
1974	784	2001	266
1975	806	2002	224
1976	784	2003	218
1977	762	2004	193
1978	765	2005	193
1979	847	2006	193
1980	911	2007	214
1981	891	2008	281
1982	941	2009	317
1983	1044	2010	366
1984	1035	2011	372
1985	1060	2012	369
1986	1007	2013	357
1987	978	2014	369
1988	949	2015	370

From Whiteshell Laboratories

2,500 AECL Reports
1,100 WNRE Reports
920 Technical Reports (TRs)
3,130 Research Company (RC) Reports

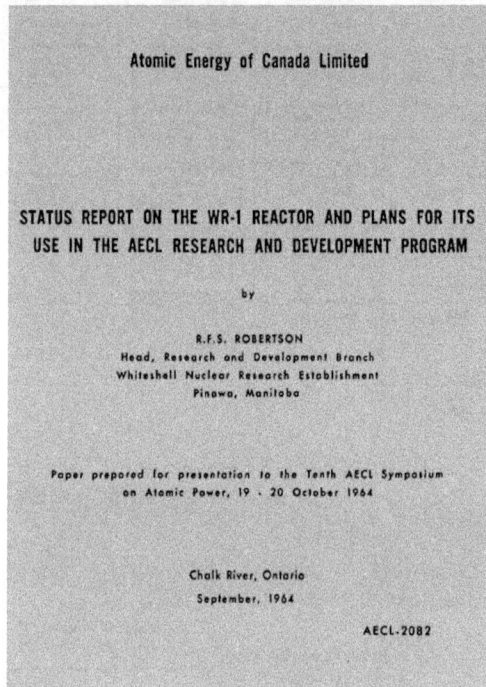

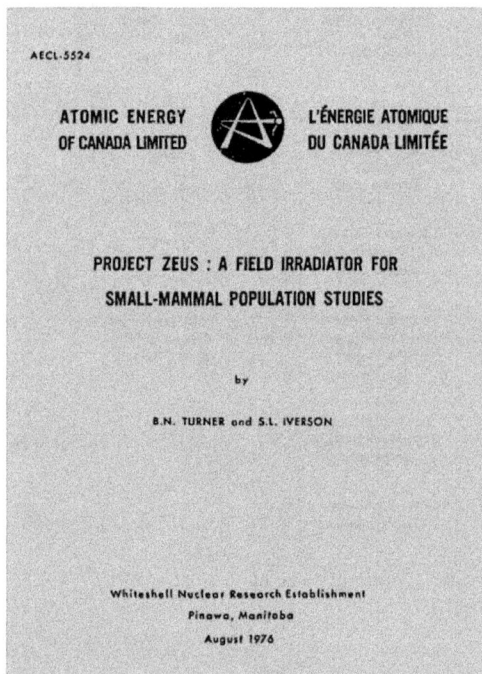

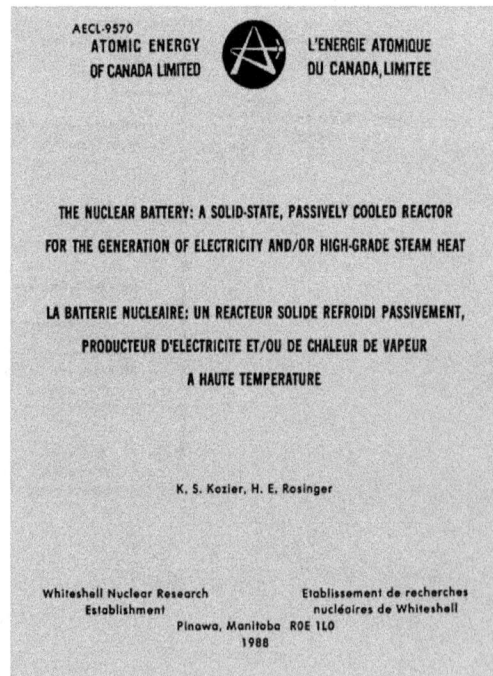

Atomic Energy of Canada Limited

STATUS REPORT ON THE WR-1 REACTOR AND PLANS FOR ITS
USE IN THE AECL RESEARCH AND DEVELOPMENT PROGRAM

by

R.F.S. ROBERTSON
Head, Research and Development Branch
Whiteshell Nuclear Research Establishment
Pinawa, Manitoba

Paper prepared for presentation to the Tenth AECL Symposium
on Atomic Power, 19 - 20 October 1964

Chalk River, Ontario
September, 1964

AECL-2082

AECL-5524

ATOMIC ENERGY
OF CANADA LIMITED
L'ÉNERGIE ATOMIQUE
DU CANADA LIMITÉE

PROJECT ZEUS : A FIELD IRRADIATOR FOR
SMALL-MAMMAL POPULATION STUDIES

by

B.N. TURNER and S.L. IVERSON

Whiteshell Nuclear Research Establishment
Pinawa, Manitoba
August 1976

AECL-9570
ATOMIC ENERGY
OF CANADA LIMITED
L'ENERGIE ATOMIQUE
DU CANADA, LIMITEE

THE NUCLEAR BATTERY: A SOLID-STATE, PASSIVELY COOLED REACTOR
FOR THE GENERATION OF ELECTRICITY AND/OR HIGH-GRADE STEAM HEAT

LA BATTERIE NUCLEAIRE: UN REACTEUR SOLIDE REFROIDI PASSIVEMENT,
PRODUCTEUR D'ELECTRICITE ET/OU DE CHALEUR DE VAPEUR
A HAUTE TEMPERATURE

K. S. Kozier, H. E. Rosinger

Whiteshell Nuclear Research
Establishment
Etablissement de recherches
nucléaires de Whiteshell
Pinawa, Manitoba R0E 1L0
1988

WNRE

TELEPHONE DIRECTORY

NOVEMBER, 1968

ATOMIC ENERGY OF CANADA LIMITED
Pinawa, Manitoba

IMPORTANT

All outside telephone calls, except those to Pinawa, are long distance and **must** be made through the Switchboard Operator.

Whiteshell Staff – 1968 Telephone Book

Abbott, M.W.
Acres, O.E.
Adams, D.E.
Aitken, J.C.
Allan, M.G.
Allwright, Mrs. E.C.
Allwright, W.A.
Anderson, F.J.
Anderson, M.B.C.
Archinuk, R.E.
Arkle, J.
Armstrong, Mrs. L.A.
Armstrong, R.K.
Ayres, W.J.

Bailey, M.G.
Baird, Miss L.C.
Baker, A.J.
Baker, Mrs. M.P.
Baker, W.J.
Balloffet, Y.
Banerjee, S.
Banham, R.B.
Banks, F.B.
Barclay, F.
Barnsdale, R.E.
Baverstock, K.F.
Baxter, D.K.
Beeskau, Miss D.G.
Behera, Dr. S.K.
Belinski, B.F.
Belinski, Mrs. P.M.
Bell, D.
Bell, Miss E.J.
Bell, L.B.M.
Bell, L.J.
Bell, T.
Bennett, C.R.
Benson, Mrs. S.M.
Benton, Mrs. A.B.
Berry, C.F.
Bird, R.L.
Birnie, T.M.
Bjornson, K.D.
Boase, D.G.
Boivin, W.A.
Booth, D.G.N.
Borsa, Dr. J.
Boulton, Dr. J.
Bowhay, G.W.
Brady, Mrs. C.M.
Brady, G.R.
Brandon, W.G.
Brandreth, M.
Bridges, H.J.
Brown, I.B.
Brown, N.R.
Brown, W.P.

Buick, Mrs. J.
Buksak, D.F.
Burnett, K.B.
Burt, G.M.
Burt, Mrs. P.
Burzynski, A.
Bushby, W.C.
Buss, D.B.
Buss, Miss L.V.
Butchart, Mrs. L.
Butler, Mrs. E.

Cadoreth, R.J.
Cafferty, J.
Cameron, C.G.
Cameron, Dr. D.J.
Campbell, D.B.
Campbell, Mrs. M.C.R.
Campbell, P.
Campbell, W.J.
Carlson, Mrs. D.D.C.
Carlson, A.G.
Carriere, D.S.
Carswell, H.
Catellier, G.
Chabluk, Miss R.J.
Chambers, D.R.
Chambers, K.W.
Chambers, M.
Chapman, Dr. J.D.
Chelack, W.S.
Chen, J.D.
Cherry, Dr. J.A.
Christoff, Mrs. E.V.G.
Chunys, W.N.
Chura, Mrs. S.L.E.
Clarke, C.F.
Clarke, Mrs. G.M.
Clarke, Mrs. M.A.
Clarke, T.E.
Clark, Mrs. V.L.
Clausen, T.D.
Clegg, L.J.
Cliché, N.P.
Coady, D.R.
Cook, G.D.
Cooper, M.D.
Cramer, D.F.
Cribbs, S.C.
Crosthwaite, J.L.
Crump, R.D.
Curie, Mrs. P.D.
Cuthbert, V.N.

Danbert, R.E.
Daun, J.K.
Davidson, H.J.
Dawson, G.P.
Decterow, V.S.

Deering, E.F.
Deneschuk, J.G.
Dent, L.N.
Dereski, W.J.
Dery, J.
Desbois, R.
Dhont, Miss J.E.
Dixon, Dr. R.S.
Dobbin, E.A.
Donnelly, M.R.
Dreger, Mrs. L.A.
Duclos, A.M.
Dugle, Dr. D.
Dugle, Dr. J.R. (Mrs.)
Dunford, W.T.
Dunford, W.W.
Dunlop, I.H.
Dunlop, T.B.
Dunn, Miss L.M.
Dutton, Dr. R.
Durand, P.
Dyck, E.
Dyck, R.W.
Dyne, Dr. P.J.

Eisenzimmer, Miss M.
Entner, D.A.

Faulkner, D.
Fay, J.G.
Fay, K.C.
Fay, Mrs. S.D.
Feicho, R.W.
Felawka, L.T.
Fenton, J.H.
Fenton, N.
Ferraro, Dr. W.C.
Finlay, Dr. B.A.
Fisher, H.L.D.
Fisher, Miss T.
Fitzsimmons, Mrs. M.E.
Fitzsimmons, T.R.
Fitzsimmons, W.D.C.
Frank, Mrs. S.A.
Futcher, A.M.

Gillespie, Dr. C.G.
Gillespie, G.E.
Gilmour, H.
Gold, J.E.
Gordon, J.B.
Gow, C.S.
Grabke, I.H.
Graham, E.V.
Grant, E.
Grant, Dr. G.R.
Greenwood, T.
Greig, D.R.
Greschuk, S.P.

Grouette, S.P.
Guthrie, J.E.

Hall, Mrs. B.J.
Hall, G.C.
Halley, S.R.
Halls, W.R.
Hamel, Dr. D.
Hammond, F.J.
Hammond, Miss K.J.
Hampton, G.
Hanley, P.T.
Hanley, Mrs. S.R.
Hansen, A.L.
Harding, J.A.
Harper, Miss L.G.
Hart, G.W.
Hart, R.G.
Hatcher, Dr. S.R.
Hawley, E.H.
Hawley, Mrs. N.J.
Hawton, J.J.
Hembroff, R.L.
Hemminger, A.
Hemmings, Dr. R.L.
Helbrecht, R.
Henschell, A.
Henschell, R.M.
Higgins, R.S.
Hiller, C.R.
Hogg, L.L.
Hollies, R.E.
Hopko, Miss B.
Horn, L.
Hornby, C.J.
Howbold, M.
Hsu, Dr. T.R.
Hunt, Miss M.
Hutchings, W.G.

Iverson, Dr. S.L.

Jabush, F.
Jackson, G.F.C.
Jackson, J.J.
Jagger, D.
Jeppeson, R.
Johnson, L.A.
Johnson, Miss L.M.
Johnstone, K.M.
Jones, B.F.
Jones, R.W.
Jones, S.
Jung, D.W.

Kabaluk, Mrs. C.A.
Kachur, M.E.
Karklin, Mrs. D.
Keefe, J.B.

Keith, V.H.
King, C.E.
King, K.L.
King, Mrs. S.
Kirkham, T.R.
Klann, P.K.
Klepatz, D.
Klepatz, Miss I.D.
Klimack, Mrs. J.L.
Kolbe, D.E.
Komadowski, B.J.
Kovacs, W.A.
Kowalchuk, Mrs. L.
Kowalchuk, W.W.
Kremers, W.
Kroeger, V.D.
Krupka, M.
Kuehl, H.R.

Lamb, T.B.
Lampard, R.G.S.
LaPorte, J.
Legiehn, M.
Leguee, G.H.
Lennox, C.G.
Lentle, Dr. B.C.
Leppaniemi, G.A.
Levesque, G.
Lidfors, E.D.
Lidstone, R.F.
Lidstone Mrs. C.
Litke, Mrs. M.I.
Litvinsky, W.
Lobban, G.J.
Loeb, R.A.
Lubitz, Miss L.C.
Lucas, R.J.
Lucas, S.D.
Lussier, R.
Luxton, C.W.
Lyon, R.B.

MacDonald, T.E.
MacFarlane, Mrs. J.C.
MacFarlane, R.
MacLean, R.H.
McCallum, C.K.
McCarthy, H.V.
McCarthy, Miss M.E.
McCooeye, D.P.
McCormac, D.D.
McDonald, E.A.
McDowell, G.
Mcghee, P.H.
McGinnis, Miss S.A.
McGovern, T.
McKay, D.A.
McKeown. W.A.
McQueen, L.J.

Macey, H.L.
Mager, B.M.
Magura, G.R.
Magura, Mrs. J.M.
Marcil, L.
Marek, Mrs. D.C.
Marks, G.L.
Martin, C.J.
Martino, J.A.
Mathers, Dr. W.G.
Mathews, P.M.
Mathews, Mrs. R.J.
Matsumoto, Mrs. C.
Matsumoto, R.
Mayor, G.H.
Mayman, S.A.
Meek, K.D.
Meeker, B.E.
Meeker, R.C.
Mehta, Dr. K.K.
Mehta, Mrs. R.
Melsness, D.J.
Meyers, G.A.
Middleton, J.E.
Mills, J.B.
Mills, R.W.
Minns, D.E.
Minton, Miss, E.J.
Minton, R.J.
Mitosinka, Miss V.A.
Mlodzinski, C.K.
Moffat, B.G.
Molnar, D.D.
Montague, R.
Montague, Mrs. R.M.
Montford, B.
Mooradian Dr. A.J.
Morgan D.N.
Morgan, W.W.
Mourareau, R.
Mowat, Miss B.B.
Moyer, R.G.
Munson, W.P.
Murphy, E.V.
Murray, D.N.
Murray, Mrs. J.P.
Murray, W.A.
Musick, C.F.
Muters, C.R.

Nayler, G.R.
Nelson, D.D.
Nerbas, D.W.
Neuls, Miss A.N.
Nicholls, H.B.

Noble, J.L.
Novakoski, R.J.

Oberick, R.F.
Ohta, M.M.
Oi, Dr. N.
Okrainec, E.
Olchowy, E.
Oldaker, I.E.
Oleskiw, E.M.
Oleskiw, Mrs. K.
Olson, H.L.
Oravec, F.G.
Ostryzniuk, Miss D.C.
Otto, Miss G.E.
Ouimet, R.
Overgaard, S.M.
Owens, J.W.

Pannell, B.J.
Patterson, D.W.
Payne, R.C.
Payne, W.E.
Peggs, Dr. I.D.
Pellow, M.D.
Penner, D.J.
Penner, G.R.
Penner, J.
Perritt, Mrs. M.M.
Perritt, T.M.
Peteleski, N.
Peterson, H.R.
Petkau, Dr. A.
Phillips, Dr. E.
Pleskach, Mrs. J.C.M.
Pleskach, J.
Pleskach, K.L.
Pleskach, S.D.
Plotnikoff, F.N.
Plotnikoff, Mrs. R.M.
Plunkett, C.H.
Pollard, L.J.
Pollock, R.W.
Porth, R.J.
Poulin, P.C.
Pouteau, Miss R.M.L.
Powaschuk, A.
Presser, A.E.
Presser, Mrs. L.D.
Prettie, R.A.
Price, A.E.
Price, D.R.
Putnam, J.M.
Puttagunta, R.V.

Quinn, M.J.

Raleigh, Dr. J.A.
Ramsay, A.
Ramsay, Mrs. A.M.
Ramsay, L.D.
Randell, R.A.
Rankin, J.
Rattai, A.C.
Reavley, Miss J.
Rector, E.S.
Reich, A.R.
Reid, Dr. W. B.
Reimer, A.
Remington, J.A.
Richter, B.
Risdahl, P.H.
Ritchie, Dr. I.G.
Robertson, A.R.
Robertson, J.J.L.
Robertson, Dr. R.F.S.
Robertson, R.W.
Rodzinsky, S.
Rogowski, A.J.
Rondeau, R.K.
Ross, R.E.
Rumak, F.T.
Ruppel, D.G.
Rylaarsdam, A.J.
Ryz, M.A.

Saarela, Mrs. P.M.
Saltvold, J.R.
Sawtzky, Dr. A.
Schatzke, D.R.
Schatzlein, E.
Schewe, S.
Schinkel, Mrs. E.
Schick, Mrs. B.J.S.
Schmidt, Mrs. C.
Schmidt, W.C.
Schultz, G.
Schultz, R.G.
Schwartz, Mrs. O.C.
Schwartz, W.J.
Schwarz, G.R.
Scott, A.R.H.
Scott, Mrs. V.A.
Serada, V.M.
Serkes, R.
Sexton, E.E.
Seymour, C.G.
Shand, G.
Shierman, G.R.

Shierman, Mrs. M.C.
Simmons, G.R.
Simmonson, L.M.
Simpson, Dr. L.A.
Sigh, Dr. A.
Singh, Dr. H. (Mrs.)
Skinner, G.
Slettede, R.
Smee, J.L.
Smiltnieks, J.
Smith, B.W.
Smith, D.W.R.
Smith, G.G.
Smith, R.M.
Snider, G.C.
Sobetski, Mrs. B.C.
Sochaski, R.O.
Sohn, K.H.
Solberg, D.M.
Solberg, Mrs. W.C.
Sopchyshyn, F.C.
Spitz, K.O.
Sprungman, K.W.
Stadnyk, Mrs. A.M.
Steffes, Miss L.M.
Steinke, D.B.
Stevens, Dr. R.
Stewart, R.B.
Strachan, W.G.
Strathdee, G.G.
Styles, R.C.
Summach, A.J.
Sutherland, D.W.
Swanson, L.M.
Swiddle, J.E.
Szekely, Dr. J.

Tabe, T.
Taylor, Mrs. O.M.
Tegart, D.R.
Theriault, F.D.
Theriault, Mrs. R.
Thexton, H.E.
Thill, D.T.
Thompson, D.A.
Thomson, R.G.
Ticknor, Mrs. J.S.
Ticknor, K.V.
Tirschmann, D.G.
Tirschmann, Mrs. J.D.
Tomchuk, R.N.
Tomlinson, M.
Trask, Miss R.
Tucker, C.W.
Turner, B.N.

Turner, D.G.
Tuxworth, Dr. R.H.
Twietmeyer, P.T.

Unger, A.E.
Unsworth, G.N.
Urbanietz, Mrs. J.E.
Urbanski, Miss G.
Usakis, A.F.

Vandergraaf, T.T.
Vanderweyde, R.D.

Walker, Dr. J.W.
Ward, I.A.
Waslyshyn, A.
Waslyshyn, Mrs. S.
Wasylenko, W.
Wasywich, Mrs. V.L.
Wazney, R.L.
Weeks, Dr. J.L.
Wegner, H.
Weissig, A.
Weissig, Miss M.R.
Whateley, T.L.
Wiebe, L.
Wiewel, R.
Wiffen, D.J.
Wiley, J.H.
Wilkins, Dr. B.J.S.
Williamson, Dr. A.
Williamson, J.P.
Wilson, T.
Witzke, K.H.
Wojciechowski, H.
Wood, Mrs. M.H.
Wood, S.G.
Woodworth, L.G.
Workman, R.W.
Wright, J.R.
Wright, M.G.
Wuschke, Mrs. D.M.
Wuschke, E.E.
Wyatt, S.G.

Yakymin, Miss O.V.
Yalden, D.S.
Yukowski, J.

Zaidman, Miss E.R.
Zerbin, A.
Zetaruk, D.G.
Zieski, Mrs. G.L.
Zink, R.J.

Whiteshell Staff – 1977 Telephone Book

Abraham, R.K.
Acres, O.E.
Agland, E.B.
Agland, R.H.
Allan, M.G.
Amundrud, D.L.
Anderson, F.J.
Anderson, M.B.C.
Anderson, P.C.
Anderstedt, E.A.
Andrews, B.W.
Armstrong, P.A.
Armstrong, R.K.
Arneson, M.C.
Arseniuk, L.A.
Arseniuk, N.S.
Artes, R.W.
Arthur, D.L.
Ayres, W.J.

Backer, H.
Bailey, M.G.
Baker, B.
Baker, R.V.
Baker, W.J.
Balness, C.C.
Balness, H.R.
Banks, F.B.
Barclay, F.B.
Barnett, P.C.
Barnsdale, R.E.
Barrie, J.N.
Baxter, D.K.
Beauchamp, C.
Bean, J.D.
Beeskau, D.G.
Belanger, V.R.
Bell, D.
Bell, D.J.
Bell, L.J.
Bell, S.J.
Bender, C.M.
Bender, H.J.
Benson, S.M.
Bera, P.C.
Berry, C.F.
Berry, M.J.
Berry, M.W.
Berry, Mike
Berube, L.L.
Bialek, R.P.
Bjornson, K.H.
Boase, D.G.
Boivin, W.A.
Booth, D.G.N.
Borgford, B.E.
Borgford, T.A.
Borsa, J.

Boulton, J.
Boulton, J.
Bowles, E.M.
Bowman, G.J.
Boyle, T.C.
Brady, G.R.
Brandon, W.G.
Bridges, H.J.
Briercliffe, R.G.
Brown, C.P.
Brown, I.B.
Brown, M.A.R.
Brown, N.A.
Brown, W.P.
Bruce, C.A.
Bruneau, D.L.
Bruneau, R.J.
Buchanan, G.V.
Bueckert, E.K.
Burnell, L.L.
Burnett, G.F.
Burnett, K.B.
Bushman, Alden
Byskal, D.
Byskal, Y.A.

Cadoreth, R.J.
Cafferty, J.
Calvert, J.L.
Cameron, D.J.
Campbell, A.B.
Campbell, P.
Campbell, W.J.
Cann, C.D.
Carefoot, L.M.
Carlson, F.C.
Carlson, L.J.F.
Carmichael, A.
Carmichael, J.A.A.
Carmichael, W.F.
Carswell, H.
Catellier, G.
Catellier, G.B.
Celli, A.
Chambers, K.W.
Chambers, M.
Chan, K.C.D.
Chapman, B.J.
Charles, P.A.
Chelack, W.S.
Chen, J.D.
Chura, L.C.
Chura, E.L.E.
Clarke, C.F.
Clarke, M.A.
Clarke, P.J.
Clegg, L.J.
Cliche, N.P.
Colborne, L.C.

Coomber, T.D.
Cooper, M.D.
Copps, T.P.
Coyne, P.M.
Cramer, D.P.
Cribbs, M.
Cribbs, S.C.
Crosthwaite, J.L.
Cutting, P.B.
Czastkiewicz, R.S.

Danbert, J.P.
Danbert, R.E.
Dancyt, N.A.
David, L.P.
Davidson, H.J.
Davis, L.E.
Davison, H.J.
Dawson, J.C.
Dawson, S.C.
Daymond, D.R.S.
Delorme, D.J.E.
Demarco, J.G.
Demoline, K.W.
Dereski, W.J.
Desbois, R.L.
Desjardins, R.F.
Declemente, S.
Dixon, R.S.
Dmytriw, M.
Dobbin, E.E.
Dobush, W.
Doern, F.E.
Donnelly, M.R.
Dooley, J.A.
Dormuth, K.W.
Douglas, P.T.
Douwes, H.S.
Dreger, L.J.
Dreger, S.M.
Dreger, V.W.
Drew, D.J.
Drynan, G.M.
Duclos, A.M.
Dugle, D.L.
Dugle, J.R.
Duncan, P.A.
Duncan, R.G.
Dunford, D.L.
Dunford, W.T.
Dunlop, I.H.
Dunlop, P.A.
Dunlop, T.B.
Dutton, R.W.
Dykstra, K.

Early, S.R.
Elcock, R.G.
Ellila, E.

Endler, H.E.
Endler, M.I.R.
Ennis, R.F.
Ewing, D.D.

Farquhar, R.W.
Faryon, T.M.
Fast, M.R.
Faulkner, D.
Faulkner, V.P.
Feicho, R.W.
Fenez, Joyce
Fenning, A.I.
Fenton, N.
Ferch, R.L.
Fiebelkorn, A.D.
Findlay, J.W.
Flemming, D.W.
Frank, S.A.
Frechette, J.L.
Frederick, J.L.
Frederick, V.P.
Funasaka, H.
Fundytus, D.

Galay, O.
Garbolinski, J.
Gardy, E.M.
Gauthier, J.R.R.
Gauthier, L.
Gauthier, N.L.
Gerwing, A.F.
Gesell, A.W.A.
Gibb, R.A.
Gillert, R.E.
Gillespie, G.E.
Gilmour, J.H.
Gilroy, S.L.
Gladys, C.
Gmitrowski, T.J.
Godin, D.P.
Gold, J.E.
Gordon, J.B.
Graham, A.
Graham, E.V.
Graham, G.H.
Graham, H.B.
Graham, H.B. Mrs.
Grant, E.E.
Grant, G.R. Dr.
Grant, G.R.
Grant, H.M.
Greenstock, C.L.
Greig, D.R.
Greschuk, S.P.
Griffith, W.
Gryseels, F.J.C.
Grzegorzewski, P.
Guthrie, J.E.

Guthrie, M.G.
Gwiazda, J.

Hachkowski, L.
Hall, G.C.
Halley, S.R.
Halliday, J.D.
Hamel, D.
Hamel, H.G.
Hamilton, A.L.
Hampton, A.K.
Hampton, G.A.
Hancox, W.T.
Hanley, P.T.
Hanley, S.R.
Hannon, E.R.
Hannon, J.D.
Harding, J.A.
Harrison, W.C.
Hart, G.W.
Hart, R.G.
Hatcher, S.R.
Haugen, H.O.
Haugen, K.A.
Havelock. F.
Havelock. P.M.
Hawkes, J.A.
Hawkins, R.J.
Hawley, E.H.
Hawley, N.J.
Haworth, N.D.
Hawryshko, W.
Hawton, J.J.
Hayamizu, Y.
Hedley, J.B.
Heidrick, T.R.
Helbrecht, R.A.
Hembroff, R.L.
Hemminger, A.
Hemminger, E.E.
Hemminger, E. Mrs.
Hemminger, J.G.
Hennig, A.
Henschell, R.M.
Henschell, R.R.
Herzog, S.G.
Hill, D.W.
Hill, F.J.
Hill, P.A.
Hillier, C.R.
Hines, C.A.J.
Hladki, D.
Hladki, N.J.
Hladun, R.
Hoard, N.E.
Hollies, R.E.
Holowchuk, M.
Holyk, A.
Honneff, P.P.

Hood, B.E.C.
Horn, L.N.
Hosein, S.M.
Howe, P.T.
Howk, R.
Hughes, F.J.
Hutchings, W.G.

Innes, D.G.L.
Iverson, S.L.
Ivison, B.J.W.

Jabush, F.W.
Jackson, J.J.
Jackson, J.M.
Jackson, K.
Jacobs, E.P.
Jacobs, I.J.
Jagannath, D.V.
James, C.W.
James, E.A.
Jansson, K.A.
Jansson, W.W.
Jarvis, A.H.
Jenks, I.H.
Johnson, H.M.
Johnston, J.E.C.
Jones, J.C.
Jones, R.W.
Jones, S.
Juhnke, D.G.
Jung, D.G.

Kabaluk, D.J.
Kaltenberger, S.
Kalupar, N.
Karklin, D.
Karklin, W.E
Karl, R.L.
Kashton, J.
Kawa, D.M.
Keefe, J.B.
Keefe, L.B.
Keith, W.W.
Kekish, S.
Kelly, J.V.
Keltie, R.J.
Kendel, J.H.
Kendel, L.L.
Kerr, J.R.
Kettles, K.M.
Kines, J.M.
King, D.E.
King, K.L.
King, S.M.
Kirkham, T.R.
Klapprat, A.G.
Klapprat, G.R.
Klapprat, W.F.
Klepatz, D.A.
Kollinger, B.J.
Komadowski, B.J.

Komadowski, E.C.
Komadowski, F.J.
Koroll, G.W.
Koroscil, G.D.
Kost, R.
Kostiuk, J.
Kovacs, D.F.
Kovacs, W.A.
Kowaluk, J.S.
Kowalchuk, G.
Kowalchuk, W.
Kozier, K.S.
Krawchuk, G.A.
Kremers, W.
Kroeger, A.
Kroeger, V.D.
Kroeker, L.R.
Krupka, M.E.
Kubish, D.E.
Kubota, K.G.
Kuehl, H.R.
Kuehl, J.H.
Kuehl, P.E.
Kuhr, E.E.
Kunz, S.
Kunzman, E.

Lachance, F.O.
Lacquement, M.D.
Lagsdine, H.
Lamb, C.E.
Lamb, T.B.
Lane, A.D.
Lange, G.L.
Laporte, J.M.
Lau, E.H.K.
Leblanc, J.C.
Ledoux, G.A.
Lee, W.
Legiehn, M.
Leitenberger, M.
Lemire, R.J.
Leneveu, D.M.
Levesque, G.L.
Lidfors, E.D.
Lidstone, R.F.
Lievarrt, P.A.
Lin, M.R.
Litvinsky, M.W.
Litvinsky, W.
Liu, D.
Liu, W.S.
Loeb, R.A.
Lofstrom, D.J.
Lofstrom, R.M.
Lopata, V.J.
Lucas, R.J.
Luczak, C.
Luke, M.O.
Lussier, R.A.J.
Lyall, L.W.

Lyon, R.B.
Lypka, R.B.

MacDonald, T.E.
MacFarlane, R.
Mackey, L.A.
MacLean, R.H.
Mager, B.M.
Magura, G.R.
Magura, J.M.
Malkoske, H.E.
Marciniak, H.F.
Marek, D.C.
Marek, R.J.
Marek, S.
Marsh, B.R.
Martino, J.A.
Marynewich, R.R.
Maslow, R.
Mathers, W.G.
Mathews, H.A.
Mathews, V.R.
Matthews, R.B.
Maxwell, J.R.
Mayman, S.A.
Mayoh, K.R.
McArthur, W.
McCallum, C.K.
McCarthy, H.V.
McCooeye, D.C.
McCormac, D.D.
McCoy, L.M.
McDonald, B.H.
McDonald, E.A.
McDowall, M.E.
McDowall, G.J.
McGinnis, E.
McIntyre, M.E.
McIntyre, N.S.
McKay, D.A.
McLean, D.R.
McLean, S.J.
McLeod, R.W.
McNaughton, J.C.
Meek, K.D.
Meeker, R.C.
Mehta, K.K.
Melin, W.P.
Mellors, J.A.
Mellors, T.J.
Mendres, R.M.
Merriman, R.J.P.
Meyer, O.H.H.
Meyer, S.
Michelfeit, J.
Middleton, J.E.
Miller, E.A.
Miller, G.G.
Miller, H.G.
Miller, M.K.
Miller, N.H.

Mills, G.A.
Mills, P.J.
Mills, R.W.
Misra, P.K.
Mitchell, B.G.
Mitchell, G.R.
Mitchell, J.H. Mrs.
Mitchell, J.H.
Moffatt, G.B.
Molinski, L.L.
Molnar, D.D.
Montague, R.
Montague, R.M.
Montgomery, J.H.
Morash, B.D.
Morgan, W.W.
Moroz, S.A.
Morris, R.W.
Moyer, R.G.
Murphy, E.V.
Murray, G.A.
Murray, W.A.
Musick, C.F.
Muzychka, E.E.
Myers, M.E.

Naaykens, D.G.T.
Nayler, G.R.
Nerbas, D.W.
Nerbas, M.A.
Newman, G.W.
Nieman, R.E.
Noble, J.F.
Noble, J.L.
Noble, L.E.
Noel, R.W.
Noel, S.E.M.
Noel, T.L.
Norek, R.J.
Norman, B.D.S.
Novakoski, R.J.
Novakowski, L.P.
Nuttall, C.M.
Nuttall, K.

Oberick, R.F.
Oberlin, G.W.
O'Connor, P.A.
Ohta, M.M.
Okrainec, E.
Olchowy, E.
Olchowy, H.J.
Oldaker, I.E.
Oravec, A.M.
Oravec, P.G.
Orlick, H.P.
Orlick, V.F.
Ott, G.H.
Ott, H.L.
Otto, R.G.
Ouimet, M.D.

Owen, D.G.

Palson, A.S.
Palson, V.G.
Pardo, Balmonte, E.
Parent, L.F.
Patterson, D.W.
Patzer, L.
Payne, R.C.
Pearson, G.P.
Pedersen, L.S.
Peggs, I.D.
Pellow, M.D.
Pellow, S.L.
Penner, G.R.
Peterson, H.R.
Peterson, M.E.
Petkau, A.
Pihulak, B.M.
Pihulak, D.G.
Pihulak, E.
Pihulak, S.
Platford, W.D.
Pleskach, J.J.
Pleskach, S.D.
Plunkett, C.L.
Plunkett, G.P.
Pollock, L.L.
Pollock, R.W.
Popple, V.L.
Porter, M.V.
Porth, E.P.
Porth, R.J.
Portman, A.
Portman, R.
Powaschuk, A.L.
Price, D.R.
Provencal, C.H.J.
Prowse, D.R.
Puls, D.G.
Puls, M.P.
Puttagunta, V.R.

Quinn, M.J.

Rajan, V.S.V.
Raleigh, D.M.
Raleigh, J.A.
Ramsay, A.
Ramsay, A.M.
Ramsay, L.D.
Randell, R.A.
Rattai, A.C.
Rattai, L.D.
Rawlings, E.B.
Reich, A.R.
Reid, R.M.
Reimer, A.
Reimer, J.
Remington, J.A.
Reschke, L.J.
Reschke, V.W.

Reykdal, L.
Richards, D.J.
Richter, B.
Rigby, G.L.
Rigby, M.I.
Robbie, D.B.
Robert, E.A.
Robertson, A.R.
Robertson, J.J.L.
Rogocki, J.R.
Rogocki, N.J.
Rogowski, A.J.
Rogowski, J.L.
Rohrig, E.J.
Rolfe, C.I.
Rolfe, T.J.
Rondeau, R.K.
Rosinger, E.L.J.
Rosinger, H.E.
Roy, N.L.
Ruehlen, R.P.C.
Rumak, A.P.
Rummery, T.E.
Ryan, S.R.
Ryz, M.A.

Saari, C.T.
Sachvie, G.H.
Sachvie, R.
Sagert, N.H.
Salvold, J.R.
Sargent, F.P.
Sargent, M.D.
Sauer, W.J.
Sawatzky, A.
Sawchuk, S.P.
Saxler, H.G.
Schankula, M.H.
Scharein, W.
Schatzlein, G.L.
Schatzlein, E.L.
Schellenberg, G.M.
Schinkel, E.R.
Schmidt, C.
Schmidt, H.K.
Schmidt, W.C.
Schultz, F.J.
Schultz, G.
Schultz, R.G.
Schwartz, O.C.
Schwartz, W.G.
Schwartz, W.J.
Seguin, S.E.E.
Seifried, J.A.
Selby, J.P.
Serkes, R.
Severson, K.L.
Sherman, E.L.
Sherman, G.R.
Shewfelt, A.J.
Shewfelt, R.S.W.

Shillinglaw, D.W.
Shoesmith, D.W.
Shorrock, L.J.
Sieg, W.
Simmons, G.R.
Simpson, L.A.
Singh, A.
Singh, H.
Sitar, W.M.
Skeet, A.G.
Skraba, T.M.
Smith, D.E.
Smith, D.W.R.
Smith, E.
Smith, F.A.
Smith, G.G.
Smith, G.W.
Smith, H.J.
Smith, I.M.
Smith, P.A.
Smith, R.M.
Snider, G.C.
Snider, G.R.
Sochaski, R.O.
Sohn, K.H.
Solberg, D.M.
Sommerfeld, H.
Soulodre, A.M.
Spencer, P.P.
Spitz, K.O.
Sprungman, K.W.
Sridhar, T.S.
Stanchell, F.W.
Stadnyk, A.M.
Stanley, F.W.
Stapleton, T.P.
Stark, D.J.
Stefaniuk, J.
Steffes, L.
Steinke, R.
Steinleitner, H.G.
Sterling, D.L.
Sterling, G.D.
Stermscheg, J.
Stewart, R.B.
Stokes, G.P.
Strathdee, G.G.
Studham, D.A.
Styles, A.
Styles, R.C.
Suchar, S.J.
Sudomlak, G.N.
Summach, A.J.
Surowiec, J.
Swanson, M.L.
Swiddle, J.E.
Szajewski, C.W.
Szekely, C.A.
Szskely, J.G.

Tabe, T.

Tamm, H.
Tammemagi, H.Y.
Tammemagi, V.
Taylor, D.V.
Taylor, P.
Taylor, R.
Terpstra, G.
Tetreault, R.L.
Tewari, P.H.
Theriault, F.D.
Theriault, R.R.
Thibault, D.H.
Thomas, D.D.
Thompson, D.A.
Thompson, D.E.
Thompson, G.W.
Thomson, R.G.
Ticknor, J.S.
Ticknor, K.V.
Ticknor, R.K.
Tiede, W.
Tighe, P.M.
Titus, W.L.
Tomchuk, B.G.
Tomchuk, R.N.
Tomlinson, M.
Too, J.J.M.
Torgerson, D.F.
Trapp, H.
Tremaine, P.R.
Truss, K.J.
Turner, B.N.
Turner, D.A.
Turner, J.G.
Turner, N.Y.
Turner, S.J.
Turpin, E.G.
Tuxworth, M.K.
Tymko, D.G.

Unger, A.E.
Unsworth, F.L.
Unsworth, G.N.
Urbanski, A.H.
Urbanski, D.L.

Vadasz, J.A.
Van Buckenhout, Y.
Vandenberghe, D.G.
Vandergraaf, T.T.
Vanderweyde, R.D.
Vankrieckenge, A.H.
Veroneau, E.L.
Voelpel, E.L.
Van Massow, R.E.
Voss, L.R.

Walker, J.F.
Wallace, G.J.
Wallace, W.R.
Walters, A.

Walters. G.A.
Walton, F.B.
Wardrop, R.T.
Wasylenko, W.D.R.
Wasywich, K.M.
Watchman, R.A.
Wayne, C.A.
Wayne, M.J.
Wazney, R.J.
Wazny, W.
Weeks, J.L.
Wegner, H.O.
Weiler, R.
Weissig, A.E.
Weisseg, E.W.
Welgan, C.J.
Wensel, G.G.
Wery, L.R.
Weselak, D.L.
Westdal, J.A.S.
Westhoven, P.J.
Whitehouse, R.P.
Wiebe, R.H.
Wiewel, A.M.
Wiggins, R.A.F.
Wikjord, A.G.
Wilkin, G.B.
Wilkins, B.J.S.
Willacy, E.
Willacy, R.
Williams, L.M.

Williamson, J.P.
Winchester, B.J.
Winchester, R.C.
Witzke, K.H.
Wojciechowski, H.
Wold, N.A.
Woligroski, M.
Woo, C.H.
Wood, M.H.
Wood, R.
Wood, S.G.
Woodbeck, W.E.
Woodworth, L.G.
Woodworth, P.E.
Wright, J.H.
Wright, M.G.
Wuschke, D.M.
Wuschke, E.E.

Young, G.J.

Zanutto, J.
Zanutto, R.L.
Zarecki, C.W.
Zechel, H.L.
Zeemel, W.G.J.
Zepp, R.A.
Zerbin, A.
Zetaruk, D.G.
Zink, J.N.
Zink, D.A.
Zirk, K.

Richard Whitehouse

Whiteshell Staff – 1986 Telephone Book

Abraham, Cathy
Abraham, Dietrich
Acres, Orville
Agland, Bob
Agland, Ethel
Aitken, Charla
Allan, Jeffery
Altstadt, Linda
Amborsky, Robert
Amiro, Brian
Amos, Eric
Amouzouvi, Francois
Anderstedt, Ted
Andres, Terry
Andrews, Bruce
Archambault, Danny
Arneson, Marvin
Arsenault, Karen
Arseniuk, Nick
Arthur, Debra
Attas, Mike
Au, Tak Keung
Auger Amiro, Connie
Augustine, Carol-Lee
Ayres, Bill
Ayers, Eric
Azzam, Edouard

Babulic, Patrick
Backer, Harry
Backer, Susan
Bailey, Bonnie
Bailey, Grant
Balness, Betty
Balness, Chris
Balness, Duane
Balness, Howard
Balness, Netta
Banks, Barrie
Banks, Bev
Barclay, Frank
Barclay, John
Barnard, John
Barnsdale, Bonnie
Barnsdale, Carman
Barnsdale, Paul
Barnsdale, Roy
Barrie, Jim
Baumgartner, Peter
Baxter, Bun
Beauchamp, Cathy
Beauchamp, Kerry
Beeching, David
Beeskau, Darlene
Behnke, Ralph
Bell, Dave
Bell, David
Bell, Shirley

Bender, Carol
Bender, Harold
Bennett, Graham
Benson, Shirley
Bera, Purna
Berry, Bill
Berry, Charlie
Berry, Michael
Berry, Mike
Berry, Stephen
Betteridge, James
Bialas, Ernie
Biddiscombe, Bud
Biebrich, Art
Bilinsky, Dennis
Bilsky, Fred
Bishop, Candy
Bishop, Jeff
Bjornson, Bud
Blais, Doug
Blais, Pat
Boase, Carol
Boase, Derek
Bochow, Heinz
Boczak, Edgar
Boivin, Bill
Booth, David
Borgford, Barb
Borgford, Dawn
Borgford, Kevin
Borgford, Thor
Borsa, Joe
Bowles, Evan
Boyer, Gordon
Boyle, Tom
Brady, Greg
Brandt, Lorraine
Bratty, Marilyn
Bratty, Ron
Bray, David
Briercliffe, Bob
Britton, Steven
Brown, Colin
Brown, Dave
Brown, Deborah
Brown, Iona
Brown, Linda
Brown, Merle
Brown, Rita
Brown, Warner
Bruce, Sandy
Bruneau, Donald
Bruneau, Lloyd
Bruneau, Marc
Bruneau, Norman
Bruneau, Robert
Buchanan, Garry
Bueckert, Ed
Burnell, Les

Burnett, Barrie
Bushman, Alden
Butterworth, Eric
Byskal, Dennis

Cadoreth, Caddy
Cadoreth, Dennis
Cadoreth, Marilyn
Cafferty, Jacquelyn
Cafferty, James
Campbell, Allan
Campbell, Phil
Campbell, Sandy
Cann, Dennis
Carlson, Floyd
Carlson, Leslie
Carlson, Paul
Carmichael, Alexander
Carreiro, Sam
Carswell, Hamilton
Carter, Amy
Carter, Dave
Carter, Ken
Carter, Margaret
Carter, Sidney
Celovsky, Andrew
Chiang, Hsu Wen
Chaboyer, Abel
Chambers, Keith
Chan, Calvin
Chan, Tin
Chapman, Barney
Chelack, William
Chen, Dennis
Chenier, Monique
Cheung, Steven
Chiang, Hsu Wen
Choi, Kwan
Choi, Yong
Choubey, Rameshwar
Chow, Alvin
Chow, Peter
Chuaqui, Claudio
Chuaqui, Offermanns Noemi
Chung, Colin
Clark, Greg
Clarke, Cyril
Clarke, Debbie
Clarke, Philip
Clegg, Leon
Cliche, Peter
Cliche, Shirley
Cooper, Brian
Cooper, Marvin
Cowdrey, Christina
Cowdrey, Eric
Cowell, Tom
Cramer, Dick

Cramer, Jan
Cribbs, Stephen
Crosthwaite, Les
Cutting, Blake
Cutting, Tanya

Daley, Steve
Dalzell, Irene
Danbert, Judy
Dancyt, Norm
Dancyt, Peggy
David, Louis
Davidson, Harley
Davis, Phil
Davison, Cliff
Daymond, Don
Decker, Donna
Delaney, Grant
Delaney, Sabine
Delannoy, Ben
Demoline, Ken
Dereski, Bill
Desbois, Ray
Desikachary, Krishna
Desjardins, Bob
Devaal, Jake
Devaal, Sheila
Dewit, William
Dheere, Douglas
Dickson, Lawrence
Dion, Therese
Dixon, Bob
Dixon, Don
Dmytriw, Metro
Dobush, Wally
Doern, Diane
Doern, Fred
Donahue, Dave
Donnelly, Brent
Donnelly, Kelli
Donnelly, Wanda
Dormuth, Ken
Drabyk, Ron
Dreger, Lloyd
Dreger, Stella
Dreger, Victor
Drew, Douglas
Drew, Patty
Drynan, Gib
Duclos, Mike
Dueck, Wilhelm
Dugle, Janet
Dunford, Bill
Dunn, Gloria
Dunn, John
Dutton, Roger
Dyck, Brian

Early, Gerry

Early, Robert
Edwards, Carole
Edwards, Kirk
Einspenner, Michael
Elcock, Rick
Ellila, Ed
Ellis, Bruce
Eschuk, Dale
Evenden, William
Everitt, Richard
Ewing, Don
Ewing, Lyn

Farquhar, Richard
Farr, Jerry
Faryon, Ted
Fast, May
Feicho, Ronald
Fenez, Joyce
Fenning, Alberta
Fenton, Nat
Fiebelkorn, Agnes
Fillion, Denis
Fillion, Ray
Findlay, Carol
Findlay, John
Fisher, Dave
Fleer, Varda
Ford, Beverly
Frech, Egon
Frechette, Agnes
Frechette, Jules

Gagawchuk, Cindy
Galay, Orest
Garbolinski, John
Gardy, Edward
Garisto, Frank
Garisto, Nava
Garroni, James
Gascoyne, Mel
Gatschuff, Helen
Gatschuff, John
Gauld, Ian
Gauthier, Annette
Gauthier, Arthur
Gauthier, Larry
Gauthier, Lynn
Geisheimer, Garry
George, Ian
George, Karen
Germain, Dave
Germain, Lawrence
Gerwing, Andy
Gesell, Alf
Gibson, Alan
Gibson, Gladys
Gillert, Bobbie
Gillert, Richard

Gillespie, George
Gilmour, Howard
Gmitrowski, Ted
Godin, Denis
Goodwin, Bruce
Goodwin, Marlene
Gorrie, John
Gowryluk, Greg
Graham, Audrey
Graham, Dennis
Graham, Glenn
Graham, Lori
Grant, George
Grant, Elna
Grant, Gary
Grant, Lyn
Grant, Yvonne
Gray, Malcolm
Greenfield, Keith
Greenstock, Clive
Greenstock, Gwen
Gregoire, Ann
Greig, Don
Grondin, Daniel
Grover, Lalit
Gryseels, Felix
Grzegorzewski, Peter
Gurela, John
Guthrie, Gladys
Guttman, Vernon

Haacke, Garry
Haines, Robert
Halley, Julie
Halley, Stewart
Halliday, Jack
Hamel, Dieter
Hamel, Henny
Hamon, Rich
Hampshire, Bob
Hampton, Allan
Hampton, Gerry
Hancox, Bill
Hanna, Bruce
Hannon, Elizabeth
Hannon, John
Harding, Jim
Harris, Debbie
Harrison, Walter
Hartrick, Darrell
Hartrick, Debbie Lee
Harvey, Keith
Hatland, Leonard
Haugen, Harry
Hawkes, Jennifer
Hawkins, Alice
Hawkins, Ray
Hawkins, Janice
Hawley, Norma
Haworth, David
Hawryshko, Bill

Hawton, Jay
Hayward, Peter
Hedley, Bruce
Heimann, Robert
Heinrich, Mary
Helbrecht, Bob
Hembroff, Betty
Hembroff, Les
Hemminger, Betty
Hemminger, Ervin
Hemminger, Jerry
Hendrickson, Don
Hennig, Art
Henrikson, Denise
Henschell, Ray
Henschell, Robin
Herman, Randy
Herzog, Stanley
Hiebert, Leonard
Hildebrand, Jack
Hildebrandt, Allan
Hill, Don
Hill, Fred
Hill, Paddy
Hillary, Beth
Hillier, David
Hillier, Jo Anne
Hiltz, Gisele
Hines, Carol
Hirst, Charles
Hirst, Shannon
Hladki, Donald
Ho, Daniel
Ho, Richard
Hoban, Michael
Hocking, Bill
Hodge, Jeannette
Hodge, Robyn
Hoffman, Janice
Hoffman, Karen
Hollies, Bob
Holloway, Art
Holyk, Andy
Honke, Glenn
Honneff, Peter
Hood, Barry
Hood, Darlene
Hood, Lana
Hosaluk, Lawrence
Hosein, Sam
Howe, Terry
Hughes, Frank
Hutchings, Bill

Ikeda, Brian
Ingham, Paul
Iverson, Stuart
Ivison, Barbara
Iwanowski, Ted

Jabush, Fred
Jackson, Jack

Jackson, Jean
Jackson, Kevin
Jacobs, Ed
Jansson, Wesley
Jarvis, Al
Jaster, Walter
Johnson, Harry
Johnson, Lawrence
Johnston, Frank
Johnston, Joanne
Johnston, John
Johnston, Laurel
Jones, Heather
Jones, Kathy
Jones, Sid
Jones, Syd
Jordan, Jim
Juhnke, David
Julien, Paul
Jung, Dieter

Kabaluk, Dennis
Kaltenberger, Mark
Kaltenberger, Silva
Kamachi, Ken
Karklin, Glen
Karl, Ray
Kautz, Brien
Kearney, David
Kearns, Tom
Keefe, Brian
Kekish, Stan
Kelly, Bud
Kendel, Lynn
Kerr, Andrew
Kerr, John
Kettles, Kim
Kines, Jim
King, Earl
King, Fraser
King, Ken
King, Sandy
King, Sharon
Kirk, Carolyn
Kirkham, Ray
Kizlik, Steve
Klapprat, Alvin
Klapprat, William
Klassen, Myrna
Klepatz, Dennis
Knight, Geoff
Knox, Margaret
Kohle, Cliff
Kolesar, Randy
Kollinger, Ben
Komadowski, Bernard
Komadowski, Edward
Komadowski, Florian
Konopetski, Frank
Koroll, Grant
Kosciuk, Andrzej

Kossman, Rainer
Kost, Robert
Kostiuk, Joe
Koudys, David
Kovacs, Dorothy
Kovacs, Karen
Kowalchuk, George
Kowalchuk, William
Kowalski, Janusz
Kowaluk, John
Kozak, Ed
Kozier, Ken
Krawchuk, Gerald
Kremers, Walter
Krishnan, Vaidyanathan
Kroeger, Aggie
Kroeger, Vic
Kroeker, Len
Krueger, Paul
Krupka, Agnes
Kubish, Don
Kubota, Kevin
Kubota, Ky
Kuhn, Terry
Kumar, Krishna
Kupferschmidt, Bill
Kuypers, Ed
Kuzyk, Greg
Kuzyk, Sharon

Laforest, Paul
Lam, Ander
Lambert, George
Lambert, Guy
Lambert, Ray
Lang, Peter
Lange, Bruce
Lange, Gerry
Laporte, John
Larocque, Alain
Lashek, Clarence
Lattanzio, Marco
Lau, Danny
Lau, Eddy
Lau, Linda
Lau, Michael
Laverock, Martha
Lavery, Bill
Lawrence, Warwick
Lawton, Todd
Lay, Doug
Leblanc, Jacques
Leblanc, Jean-Claude
Ledoux, George
Ledoyen, Norman
Lee, Albert
Legall, Anita
Legiehn, Manfred
Legiehn, Olga
Leitch, Brian
Lemire, Alberta

Lemoine, Andre
Leneveu, Dennis
Lesage, Holly
Leszkowicz, John
Lidfors, Dale
Lidstone, Bob
Lim, George
Liner, Yakov
Lisle, David
Litke, Bill
Litke, Cynthia
Litke, David
Litke, Paul
Litvinsky, Walter
Lobreau, Anya
Locke, Roy
Lodge, Bryan
Lodge, Cindy
Loeb, Ron
Loesewitz, Wolfgang
Lofstrom, Dennis
Lofstrom, Rosalie
Lopata, Vince
Lopez, Raul
Loudon, Roy
Lucas, Pat
Lucas, Roy
Luczak, Chester
Lussier, Robert
Lutz, Stephanie

MacDonald, Bill
MacDonald, Tom
MacFarlane, Roy
MacLean, Glen
MacLean, Ron
MacMillan, John
MacRae, Gilbert
Maddock, Willard
Mager, Budd
Maguire, Kevin
Magura, Gord
Magura, Judith
Maier, Kathryn
Main, Varda
Mains, Anne
Malkoske, Harold
Mallory, James
Mamrocha, Eugene
Mamrocha, Randal
Manchulenco, Debra
Manchulenco, Jim
Marciniak, Henry
Marciniak, Mary
Marek, Stan
Martino, Jerry
Marusyk, Brad
Maslow, Ron
Mathers, Bill
Mathew, Aleykutty
Mathew, Mani

Spinney, Hugh
Spitz, Klaus
Sitz, Mary
Sprungman, Kurt
Sridhar, Sampat
St. Denis, Bonnie
Stabner, Tim
Stanchell, Frank
Stanley, Wayne
Stark, Dennis
Stark, Ingrid
Stefaniuk, John
Steiner, Vern
Steinke, Rudy
Steinleitner, Hans
Stelko, Larry
Stepanik, Terry
Stephens, Michael
Sterling, Gary
Sterling, Lloyd
Stokes, Gary
Stone, Bob
Stroes Gascoyne, Simcha
Studham, Dave
Styles, Anne
Styles, Roy
Sullivan, Pat
Sunder, Sham
Suski, Richard
Sveinson, Sheldon
Swanson, Lorne
Swanson, Tim
Swiddle, Jack
Syring, John
Szajewski, Casmir
Sze, Yukeung
Szekely, Carol

Szskely, Joseph
Tabe, Tom
Tamm, Heiki
Tamm, Judy
Taylor, Ann
Taylor, Douglas
Taylor, Peter
Taylor, Olive
Taylor, Roy
Tennankore, Kanan
Terpstra, Gerrit
Tetreault, Rosaire
Thephasone, Chum
Thephasone, Kham
Theunissen, Hank
Thibault, Dennis
Thomas, Alex
Thomas, Bryan
Thomas, Dennis
Thompson, Paul
Thomson, Dick
Thomson, Maryann
Thorne, Garry
Ticknor, Joan
Ticknor, Ken
Ticknor, Roy
Tiede, Wilson
Tirschmann, Art
Torac, Jean
Tomlinson, Mike
Torgerson, David
Towle, Rick
Tracy, Brian
Tran, Francis
Trapp, Harold
Trudeau, Donald
Truss, Kent
Tufford, Olivia

Turner, Dorothy
Turner, Jack
Turner, Sharon
Tutiah, Marvis
Tuxworth, Martha
Tymko, David
Unger, Abram
Ungurian, Mark
Urbanski, Ken
Vadasz, Joseph
Van Buckenhout, Yvonne
Vance, Lou
Vandergraaf, Chuck
Vanderweyde, Doug
Vatsya, Shiv
Veilleux, Michael
Velie, Delores
Veroneau, Ed
Vijayan, Sivaraman
Vik, Ole
Vikis, Andreas
Vilks, Peter
Vincett, Bert
Voelpel, Elmer
Van Massow, Rudi
Voss, Lloyd
Walker, James
Wallace, Dianne
Wallace, Gary
Wallace, Sue
Wallace, Ward
Walters, Agnes
Walters, Gerry
Walton, Carol
Walton, Frank

Wan, Pak Tai
Wasylenko, Wayne
Waters, Howard
Watson, Rogan
Watson, Shona
Wayne, Mike
Wazney, Ray
Weber, Ray
Wedgwood, John
Weeks, John
Wegner, Harold
Welz, Nadine
Wensel, Gerry
Werner, Bob
Weselak, Debbie
Westdal, John
Westhoven, Peter
Whitaker, Sidney
White, Andrew
Whitehouse, Richard
Wiewel, Tony
Wikjord, Alf
Wilgosh, Brenda
Wilgosh, Bruce
Wilgosh, Glen
Wilkin, Bruce
Wilkin, Debra
Wilkins, Brian
Wilkins, Sylvia
Williams, Juanita
Williams, Len
Williams, Mike
Williamson, Pat
Wilson, Gisele
Winchester, Joan
Winchester, Richard
Witzke, Ken
Wold, Curtis

Wold, Norman
Woo, Chung
Wood, Donald
Wood, Lucille
Wood, Stanley
Woodbeck, Bill
Woodbeck, Kathy
Woodcock, David
Woodworth, Len
Woodworth, Pat
Wooster, Scott
Wren, Clara
Wren, Dave
Wright, Esther
Wright, Fiona
Wright, Hilton
Wright, Mike
Wuschke, Donna
Wuschke, Ed
Ylonen, Louise
Young, Allan
Young, Gary
Younger Lewis, Douglas
Yuen, Peter
Yule-Charles, Pam
Yurkowski, Ron
Yurkowski, Teena
Zach, Marg
Zach, Reto
Zarecki, Cliff
Zechel, Jerry
Zerbin, Ad
Zetaruk, Don
Zirk, Ken

Duane Balness and Keith Reid

Sandra Ramsay, Noemi Chuaqui-Offermanns and Claudio Chuaqui

Whiteshell Staff – 1994 Telephone Book

Abraham, Cathy
Abraham, Dietrich
Acres, Orville
Aitken, Charla
Alexander, Gaille
Allan, Colin
Allan, Jeffery
Altstadt, Linda
Amiro, Brian
Amos, Eric
Amouzouvi, Francois
Anderstedt, Ted
Andres, Terry
Andrews, Bruce
Archambault, Danny
Archambault, Tara
Arneson, Colombe
Arneson, Marvin
Arsenault, Karen
Arthur, Debra
Ates, Yusuf
Attas, Mike
Au, Doug
Augustine, Carol-Lee
Aydemir, Nusret
Ayres, Eric
Ayres, Heather

Bachinski, Donald
Backer, Harry
Backer, Rick
Backer, Susan
Bailey, Bonnie
Bailey, Grant
Balachandar, Ram
Ball, Joanne
Balness, Betty
Balness, Chris
Balness, Duane
Balness, Howard
Balness, Netta
Barnard, John
Barnsdale, Adonna
Barnsdale, Paul
Barnsdale, Thomas
Baspaly, Len
Baumgartner, Peter
Baxter, Bun
Beauchamp, Cathy
Beauchamp, Kerry
Bedford, James
Beeching, David
Behnke, Edith
Behnke, Ralph
Bell, David
Bender, Harold

Benson, Shirley
Bentley, Terry
Bera, Purna
Beres, Tomas
Berry, Charlie
Berry, Mike
Berry, Stephen
Betteridge, James
Beuthe, Thomas
Bialas, Ernie
Bilinsky, Dennis
Bilsky, Fred
Bird, Glen
Bisaillon, Philip
Bishop, Candy
Bishop, Jeff
Bisko, Gary
Bjornson, Kenneth
Blom, Diane
Blowers, Danny
Boczak, Diane
Boczak, Edgar
Boivin, Bill
Boivin, Johanne
Bonekamp, Sandra
Borgford, Barb
Borgford, Kevin
Borsa, Joe
Bowles, Evan
Boyle, Shirley
Boyle, Tom
Brady, Greg
Bratty, Marilyn
Bratty, Ron
Brennan, Yvonne
Breton, Danny
Brincheski, Glen
Brincheski, Margaret
Broadfoot, Myrna
Broadfoot, Roderick
Brown, Anton
Brown, Colin
Brown, Debbie
Brown, Linda
Brown, Merle
Brown, Morgan
Bruneau, Crystal
Bruneau, Don
Bruneau, Lloyd
Bruneau, Marc
Bruneau, Norman
Bruneau, Robert
Buchanan, Garry
Bueckert, Ed
Buell, John
Burgoyne, Donna
Burnell, Les
Burnett, Barrie

Burns, Blayne
Burns, Rose
Bush, Angel
Bushman, Alden
Butterworth, Eric
Byskal, Dennis

Cafferty, James
Cafferty, Lesa
Campbell, Allan
Campbell, Sandy
Cann, Dennis
Carlson, Floyd
Carlson, Leslie
Carlson, Paul
Carter, Dave
Carter, Sidney
Champagne, Kimberley
Chan, Calvin
Chandler, Neil
Chapman, Barney
Chen, Dennis
Chernis, Peter
Chiang, Hsu Wen
Choubey, Rameshwar
Chow, Peter
Chung, Colin
Chung, Minda
Clarke, Alicia
Clarke, Cyril
Clarke, David
Clarke, James
Clarke, Philip
Clegg, Leon
Coleman, John
Colotelo, Catherine
Cooper, Brian
Corbett, Brian
Cote, Sylvie
Cowell, Tom
Cramer, Jan
Cribbs, Mary Ann
Cribbs, Stephen
Crognali, Rosa
Crosthwaite, Christine
Crosthwaite, Les
Crosthwaite, Paul
Cutting, Blake
Cutting, Tanya

Dalzell, Shannon
Dancyt, Norm
David, Louis
Davison, Cliff
Daymond, Don
Daymond, Louise
Delaney, Grant
Delannoy, Ben

Demoline, Ken
Desikachary, Krishna
Desjardins, Bob
Desrochers-Alpers, Lise
Dixon, David
Dmytriw, Metro
Dobush, Wally
Doern, Diane
Doern, Fred
Donahue, Dave
Donnelly, Brent
Donnelly, Jim
Donnelly, Wanda
Dooley, Fay
Dormuth, Alice
Dormuth, Ken
Drabyk, Anita
Drabyk, Ron
Drake, Myrna
Dreger, Lloyd
Dreger, Stella
Drew, Douglas
Drew, Patty
Driver, Paul
Drynan, Gib
Drynan, Myles
Duclos, Mike
Dueck, Willie
Duna, Vienafe
Dutka, Arthur
Dutton, Roger
Dyck, Brian
Dyck, Gary
Dykstra, Sharon
Dzik, Ed

Early, Gerry
Early, Robert
Edwards, Carole
Edwards, Kirk
Ejeckam, Reg
Elcock, Christopher
Elcock, Judith
Ellis, Bruce
Ellis, Ron
Endler, Harvey
Eschuk, Dale
Evans, Derri
Evans, Lorna
Evenden, William
Everitt, Richard
Ewing, Don
Ewing, Lyn

Farquhar, Richard
Farr, Jerry
Farrell, Wanda
Feicho, Ronald

Fenton, Nat
Fiebelkorn, Agnes
Fillion, Ray
Findlay, Carol
Findlay, John
Flett, Debbie
Fraser, Rhea
Frechette, Agnes
Frost, Laurie

Gadallah, Lawrence
Gann, Paul
Gardy, Edward
Garneau, Luc
Garroni, James
Gascoyne, Mel
Gauld, Ian
Gauthier, Annette
Gehon, Douglas
George, Ian
Germain, Dave
Germain, Lawrence
Gerwing, Andy
Gesell, Alf
Gesell, Trevor
Ghuttora, Sukee
Gibson, George
Gillespie, George
Gillespie, Phyllis
Glowa, Glenn
Gilmour, Howard
Gmiterek, Teresa
Godin, Denis
Godin, Gill
Good, Daniel
Goodman, Troy
Goodwin, Bruce
Goodwin, Marlene
Gorrie, John
Gowryluk, Geg
Graham, Dennis
Graham, Lori
Gray, Barbara
Gray, Malcolm
Greber, Mary
Greenfield, Keith
Greig, Don
Griffault, Lise
Grzegorzewski, Peter
Guerrero, Abraham
Gurela, John
Guttman, Grace

Haacke, Garry
Haacke, Kathleen
Hajas, Wayne
Halley, Julie
Halley, Stewart

Hamon, Connie
Hampshire, Bob
Hampshire, Joan
Hampshire, Sheri
Haner, Wayne
Hanna, Bruce
Hanna, Kathryn
Hannon, John
Hansen, Ethel
Harding, Jeffrey
Harding Shirley
Harris, Debbie
Harrison, Ruth
Harrison, Walter
Harvey, Kay
Harvey, Keith
Hatland, April
Hatland, Leonard
Haugen, Kirk
Hawkins, Janice
Haworth, David
Hawryshko, Bill
Hawton, Jay
Hayles, John
Hayward, Peter
Heckert, Donna
Hedley, Bruce
Helbrecht, Bob
Hembroff, Les
Hemminger, Ervin
Hemminger, Jerry
Hennig, Art
Henschell, Eileen
Henschell, Ray
Henschell, Robin
Herman, Randy
Hiebert, Irene
Hiebert, Leonard
Hildebrand, Jack
Hiltz, Gisele
Himbeault, Donald
Hines, George
Hladki, Donald
Hladki, Jimmy
Hladki, Norma
Hladun, Bonnie
Hnatiw, David
Hnatiw, Joan
Hoban, Michael
Hocking, Bill
Hocking, Margaret
Hodge, Jeffrey
Hoffman, Dana
Hoffman, Janice
Hogeveen Ungurian, Mark
Honke, Glenn
Hood, Barry
Hood, Darlene
Hosein, Rose
Hosein, Sam

Howe, Terry
Howk, Susan
Hughes, Frank
Hume, Harold

Ikeda, Brian
Ingham, Paul
Iverson, Stuart
Iwanowski, Ted

Jackson, Kevin
Jacobs, Ed
Jansson, Wesley
Jarvis, Al
Jobe, David
Johnson, Ellen
Johnson, Harry
Johnson, Lawrence
Johnson, Pat
Johnston, Frank
Johnston, Joanne
Johnston, Laurel
Jones, Syd
Jordan, Jim
Juhnke, David
Jung, Dieter

Kaatz, Richard
Kaatz, Susan
Kabaluk, Dennis
Kaltenberger, Mark
Karklin, Glen
Karl, Ray
Kashton, David
Kasprick, Gregory
Kawa, Dennis
Kearns, Judy
Kearns, Tom
Keefe, Brian
Keeling Stephen
Keith, Darlene
Keith, Shawn
Kekish, Stan
Kellendonk, Cindy
Kendel, Lynn
Kerr, Andrew
Kerr, John
Kesanen, Eric
Kettles, Kim
Kettlewell, David
Khair, Khairy
Khan, Nazeer
King, Earl
King, Fraser
King, Ken
King, Sandra
King, Sharon
Kingsland, Laurie
Kirk, Carolyn
Kitson, Charles
Kizlik, Steve
Klepatz, Dennis

Klepatz, Linda
Knight, Geoff
Knox, Margaret
Kohle, Cliff
Kolar, Miroslav
Kolesar, Randy
Kollinger, Ben
Komadowski, Bernard
Komadowski, Edward
Komadowski, Florian
Komadowski, Terry
Koroll, Grant
Kosciuk, Andrzej
Kossman, Maureen
Kost, Robert
Kostiuk, Joe
Kovacs, Karen
Kovari, Marilyn
Kowalchuk, George
Kowalchuk, William
Kowalski, Janusz
Kowaluk, John
Kozak, Ed
Kozier, Ken
Krause, Ken
Krawchuk, Gerald
Kremers, Walter
Kroeger, Aggie
Kroeger, Vic
Kroeker, Len
Krueger, Paul
Kruk, Stan
Kubish, Don
Kufflick, Dori
Kukurudz, Joanne
Kumar, Krishna
Kupferschmidt, Bill
Kuzyk, Greg
Kuzyk, Sharon

Lambert, Guy
Lambert, Ray
Lange, Gerry
Laporte, John
Lashek, Clarence
Lau, Danny
Laurin, Suzanne
Lavac, Marc
Laverock, Martha
Lavoie, Connie
Lavoie, Omer
Leblanc, Jacques
Leblanc, Jean-Claude
Ledoux, George
Ledoyen, Norman
Lee, Albert
Legiehn, Manfred
Legiehn, Olga
Lei, Quanmin
Leitch, Brian
Leitch, Gary

Leneveu, Dennis
Liblong, Steve
Lidfors, Alanna
Lidstone, Bob
Lim, George
Lim, Monica
Lincoln, Jeff
Lincoln, Raquel
Lindquist, Eric
Litke, Cynthia
Litke, David
Litwin, Keith
Lodge, Cindy
Lodha, Ganpat
Loeb, Ron
Lofstrom, Dennis
Lofstrom, Rosalie
Lopata, Vince
Lortie, Bonnie
Lucht, Lisa
Luke, Michael
Lussier, Robert

MacDonald, Bill
MacDonald, Colin
MacDonald, Tom
MacDonald, Maureen
MacFarlane, Roy
MacLean, Glen
MacMillan, Jim
Mager, Budd
Maguire, Kevin
Magura, Gord
Malkoske, Harold
Mallory, James
Mamrocha, Eugene
Mamrocha, Randal
Manchulenco, Jim
Markham, Gord
Marciniak, Mary
Marlowe, Steve
Marohn, Susan
Martin, Derek
Martin, Heather
Marek, Stan
Martino, Jason
Maslow, Ron
Mathew, Mani
Mathews, Sandy
Matthews, Virgina
Mauthe, Karen
Mauthe, Ralph
May, Dawn
Mayoh, Keith
McArthur, Loreen
McAuley, Gloria
McCalder, Terry
McCallum, Ken
McCamis, Robert
McClinton, Bruce
McConnell, Jodi

McCooeye, Danny
McCooeye, Pat
McCoy, Lydia
McCrank, Glen
McDonald, Eve
McDonnell, Frank
McDougall, Tom
McDowall, Barb
McDowall, Bonnie
McDowall, Marg
McDowall, Sandy
McEachern, Rod
McFarlane, Joanna
McGarvey, Bryce
McGee, Glen
McGregor, Robert
McIlwain, Alan
McIlwain, Heidi
McKee, Rick
McLean, Don
McLean, Robert
McLeod, Wayne
McMurry, Jude
Meade, Alice
Meads, Cheryl
Meads, Donald
Meek, Ken
Mehta, Kishor
Mellors, Joyce
Mellors, Trevor
Melnyk, Alan
Melnyk, Ted
Merritt, John
Meyer, Larry
Meyer, Oswald
Meyer, Siegrun
Meyers, Leslie
Mezyk, Stephen
Middleton, Jack
Miller, Hans
Miller, Maureen
Miller, Neil
Miller, Susan
Mills, Pat
Moir, Deborah
Molinski, Len
Moltyaner, Freda
Monaster, Roberta
Montgomery, George
Moore, Christopher
Morash, Brian
Moreau, Andy
Moreau, Mark
Morrish, Donna
Motycka, Michael
Moyer, Ralph
Muir, Ian
Mulpuru, Sam
Murphy, Chuck
Murphy, Pearl
Murray, Gordon

Whiteshell Laboratories

Usackis, Andy

Vandergraaf, Chuck
Veilleux, Julie
Veilleux, Michael
Veroneau, Ed
Veroneau-Jansson, Terri
Vikis, Andreas
Vilks, Peter
Voelpel, Elmer
Voss, Lloyd

Walker, Bradley
Wallace, David
Wallace, Gary
Wallace, Sue
Wallace, Ward
Walters, Agnes
Walton, Carol
Walton, Frank
Wan, Alan
Warenko, Raymond

Wasylenko, Wayne
Wasywich, Ken
Waters, Howard
Watson, Angela
Watson, Rogan
Wayne, Mike
Wazney, Ken
Weber, Ray
Wedgewood, Joanne
Wedgwood, John
Wegner, Harold
Welz, Nadine
Wensel, Gerry
Westdal, John
Whitaker, Sidney
Whitehouse, David
Whitehouse, Richard

Whittaker, Alan
Wiewel, Tony
Wikjord, Alf
Wilcox, Brian
Wilgosh, Brenda
Wilgosh, Bruce
Wilgosh, Glen
Wilken, Dorothy
Wilkin, Bruce
Wilkins, Brian
Wilkins, Sylvia
Wilson, Gisele
Winchester, Joan
Winchester, Richard
Witzke, Ken
Wojciechowski, Laverne
Wold, Kathy

Wold, Norman
Woo, Chung
Wood, Donald
Wood, Lucille
Wood, Stanley
Woodbeck, Kathy
Woodcock, David
Woodworth, Len
Wooster, Scott
Worona, Heather
Worona, Shannon
Wren, Clara
Wren, Dave
Wright, Esther
Wright, Mike
Wuschke, Donna

Young, Allan
Young, Gary
Young, Patricia

Zach, Marg
Zach, Reto
Zahorodny, Janice
Zarecki, Cliff
Zerbin, Ad
Zetaruk, Don
Zhuang, Yahui
Zink, Rodney
Zirk, Denyse
Zirk, Ken
Zukic, Nermin

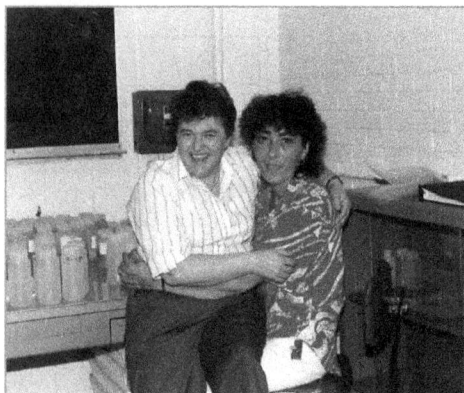

Merle Brown and Ela Rochon

Larry Shewchuk

Les Burnell

Charlie Berry

Joanne Ball and Bill Kupferschmidt

Whiteshell Staff – 2013 Telephone Book

Abercrombie, Corey
Abraham, Richard
Adams, Leah
Aitkenhead, Christel
Akhtar, Jawed
Akpan, Uduak
Alam, Shamsul
Alarie, Kevin
Alex, Artur
Alhalabi, Fahed
Amos, Paul
Antymis, Dean
Austin, David
Avanthay, Kim
Avanthay, Simon
Ayres, Eric

Baader, Michael
Bablecos, Bill
Bachman, Lori
Backstrom, Joseph
Bakosti, Bellamy
Balness, Duane
Banack, Jason
Bannish, Daniel
Barager, Kevin
Baranov, Artem
Barnett, Gina
Barnsdale, Adonna
Barnsdale, Paul
Barrios, James
Barron, Tim
Benger, George A.
Bernauer, Tyler
Berry, Michael
Berry, Stephen
Berry, Todd
Betteridge, James
Bialas, Ernie
Bilinsky, Ryan
Bilkoski, Kenneth W.
Blowers, Christine
Blowers, Danny
Bodley, Sue
Bonekamp, Jack
Borgford, Kevin
Borton, Scott
Boulet, Denis R.
Boyle, Michael
Breton, Dustin
Breton, Tim
Brincheski, Glen
Bruneau, Toni
Buchanan, Gerard
Buell, John
Bukoski, Jeffery
Bush, Cory
Bush, Kelly

Byskal, Dennis

Carolan, Josh
Caron, Allan
Caya, Andre
Chevrefils, Rene
Chrusch, Kyle
Chshyolkova, Tatyana
Clarke, Kevin
Clarke, Philip
Clayton, Brendon
Conroy, Karen
Corrigan, Shawn
Coss, Jonathon
Cowan, Barrie
Crognali, Rosa
Crosthwaite, Paul
Cummer, Jody
Cure, Nicole
Cyncora, Jayson

David, Adam
Davis, Charene
Dearsley, Colin
Demers, David
Desbois, Ryan
Donald, Bill
Donnelly, Brian
Donnelly, Tavis
Dooley, Fay
Drabyk, Anita
Drivas, Alexandros
Dueck, Sheldon
Dueck, Steven L.
Duff, Kevin
Dugard, Jeff
Duncan, Cameron
Dunlop, Dwight
Dupont, Mike
Durand, Ray
Dykstra, Quinn

Edwards, Carole
Edworthy, Jamie
Elkin, Cindy
Ellert, Andrew
Emond, Lisa
Enns, Mike
Erickson, Trish
Escasura, Enrique
Eyolfson, Neil

Falk, Jason
Favreau, Brian
Felushko, Kent
Fermin, Albert T.
Fillion, David
Fitzmaurice, Ron

Flood, Allan
Forbes, Kim
Fowler, Brian
Frith, Braden
Furnish, Mike

Gagnon, Dan
Gagnon, Noeme
Galeschuk, Dravel
Gamley, Sean
Gartner, Vojko
Gerardy, James
Germaine, Doug
Gervais, Joanne
Gesell, Trevor
Giesbrecht, Shawn
Gladish, Eric
Gmiterek, Teresa
Graham, Andrew
Graham, Curtis T.
Graham, Lori
Grant, Doug
Grzegorzewski, Denny
Grzegorzewski, Julia
Gunn, Kirk
Gwozdz, Marcus

Harding, Jeff
Haugen, Kirk
Hayter, Jeff
Hebb, Mike
Hebert, Donald
Helbrecht, Kevin
Hembroff, Mark
Henschell, Perry
Hessian, Cindy
Hirst, Sheri
Hladki, Jimmy
Hlady, Grant
Hoey, Len
Hogeveen Ungurian, Mark
Howlett, Don

Ingeberg, Luke
Ivory, Zeke

Jack, Neil
Jackson, Kevin
Jacob, Manny
Jansson, Brenda
Jenkins, Trefor
Jones, Syd

Kaatz, Rick
Kaltenberger, Mark
Kaminski, Jesse
Keith, Darlene

Keith, Shawn
Kennedy, Troy
Khan, Syed
Kim, Chang-Seok
Kitson, Charles
Klees, Jeff
Kohle, Clifford
Kolesar, Randy
Koroll, Grant
Koroscil, Bryan
Kost, Don
Koak, Ed
Kroker, Ken
Kryschuk, Chantelle
Kryschuk, David
Kuhl, Sherri
Kuzminski, Peter
Kuzyk, Steve

La Rue, Nicole
Lafreniere, Alex
Laliberte, Daniel
Lam, Philip
Langtry, Jesse
Larsen, Kevin
Laurin, Suzanne
Lavallee, Wendy
Lavallee, William
Laverock, Martha
Lavoie, Omer
Leake, Lorraine
Leishman, Connie
Leonard, Wayne
Liddiard, Riley T
Lim, Ric
Lin, Lenny
Lindsay, Fiona
Litke, Jared
Liu, Fuping
Lodge, Ken
Lowen, Ed
Luke, Adriana

Maguire, Kevin
Malek, Kamil
Mallory, James
Mamrocha, Randal
Markham, Gord
Martini, Bruce
Martino, Jason
Matheson, Sandi
May, Doug
McDonald, Fiona
McLean, Robert
McLean, Suzanne
McMillan, Steven
McMurren, Devin
Meadwell, Eric

Meek, Todd
Meeker, Pamela
Melillo, Joe
Michaluk, Craig
Milhausen, Dawn
Miller, Evan
Miller, Jeff
Miller, Neil
Mitchell, Daniel
Mitchell, Melanie
Mlodzinski, Anna
Molinski, Tammy
Mulhall, Darren
Mynhardt, Sunette
Mysak, Gary
Myshkowsky, Dan

Nally, Ritchie
Newton, Chad

Onagi, Dwayne
Orbanski, Bruce
Otto, Lorne
Oversby, Craig

Pachkowsky, Cameron
Pachkowsky, Grant
Pachkowsky, Lori
Palchewich, Darren
Pannu, Balraj
Panson, Jamie
Papineau, Noel
Pargeter, Pat
Parisian, Sherry
Parrott, Stuart
Patel, Dipak
Pawluk, Patricia
Peitsch, Neil
Petrea, Adrian
Pisiak, Gord
Plew, Andrew
Podaima, Glenn
Podolsky, Corey
Pommer, Vernon
Popiel, Trevor
Powney, Jim
Predinchuk, Ryan
Priyanto Putro, Deni
Prymak, Don
Purcell, Ed
Purtill, Mitch

Radoiu, Mike
Radyastuti, Amina
Ramsay, Frazer
Rasheed, Sana
Rasmussen, Leonard
Recksiedler, Todd

Whiteshell Laboratories

Redmond, Angela
Regula, Neil
Reid, Duncan
Reidke, David
Reimer, Eleanor
Reimer, Kim
Reimer, Mike
Reimer, Terry
Renard, Edna
Ridgway, Randall
Ritchot, Richelle
Roche, Tom
Rogers, Ainslie
Rogers, Kevin D.
Rollins, Gary
Rosentreter, Michelle
Rosentreter, Tim
Rowlands, Jenn
Ruta, Jodi

Sackvie, Ken
Salter, Rob
Scharer, George
Schatkowsky, Wendell
Schmidt, Paul
Schuhmann, Siggi
Schultz, Clive
Schultz, Neal
Shewchuk, Wayne
Sieg, Joanne
Singbeil, Celeste
Smith, Mike
Smyrski, Sherri
Sokoloski, Scott W
Sorenson, Chris
Specaluk, Gavin
Stadnyk, Trevor
Stefaniuk, Barry
Stefaniuk, Janet
Stefaniuk, Shannon
Steiner, Vern
Stelko, Larry
Stelko, Melissa
Stelmack, Curt
Stepanik, Terry
Stephenson, Kate
Struss, Ashlee
Strymbiski, Jeff
Suresh, Manu
Swaenepoel, Jason
Swaenepoel, Shannon
Swain, Laurissa
Swartz, Randall
Sytnick, Callan

Tapia, Elen
Taylor, Angela
Taylor, Ben
Taylor, Louise
Thiessen, Kati
Thomas, Bryan
Thomas, Mike

Thomas, Trevor
Thompson, Paul
Thomson, Cameron
Ticknor, Kevin
Tiede, Myrna
Tiede, Robert
Tresidder, Steven
Tymko, Kelly L

Vanden Berg, Rick
Veilleux, Clinton
Vilks, Peter
Voth, Alfie

Wallach, Paul
Warenko, Raymond
Weselak, Nolan
Whiting, Gregory
Wiebe, Carly
Wiebe, Robert
Wiens, Walter
Wiklund, Kyle
Wilcox, Alanna
Wilcox, Brian
Wilcox, C. Brian
Wilgosh, Brenda
Wilgosh, Bruce
Wilgosh, Glen
Wilgosh, Monica
Wilson, Cory
Wilson, Leslie
Witoski, Len
Witte, Arnold
Wojciechowski, Laverne
Wojcik, Donald
Wood, Darryl
Wood, Garry
Woodcock, David
Worona, Anthony
Worona, Randy
Worona, Shannon
Wu, Jason

Xu, Daisy

Yarish, Jordan
Yassin, Nidal
Yule, Pamela

Zacharias, Adam
Zelinsky, Derrick
Zhou, Eric
Zieske, Kevin
Zimmer, Jerrad
Zirk, Kyle

Randy Worona

Glenn Podaima

Whiteshell History Committee

Chris Saunders

Colin Allan

Mike Attas

Joseph Borsa

Bob Dixon

Stu Iverson

Grant Koroll

Stewart McIntyre

Steve Mihok

Robert Pollock

Peter Sargent

Leonard Simpson

Ray Sochaski

Alf Wikjord

The Whiteshell History Committee is responsible for the writing of this book. All information for the book was collected from internal AECL reports, public documents and the memories of Whiteshell employees. Many of the committee members also provided articles, photographs and hours of editing services. Thank-you

www.ingramcontent.com/pod-product-compliance
Lightning Source LLC
Chambersburg PA
CBHW051214200326
41519CB00025B/7111